T0290960

GLOBAL NAVIGATION SATELLITE SYSTEM MONITORING OF THE ATMOSPHERE

GLOBAL NAVIGATION SATELLITE SYSTEM MONITORING OF THE ATMOSPHERE

GUERGANA GUEROVA
Associate Professor, Department of Meteorology and Geophysics, Sofia University "St. Kliment Ohridski", Sofia, Bulgaria

TZVETAN SIMEONOV
Research scientist, GRUAN Lead Center, Deutscher Wetterdienst (DWD), Lindenberg, Germany

ELSEVIER

Elsevier
Radarweg 29, PO Box 211, 1000 AE Amsterdam, Netherlands
The Boulevard, Langford Lane, Kidlington, Oxford OX5 1GB, United Kingdom
50 Hampshire Street, 5th Floor, Cambridge, MA 02139, United States

Library of Congress Cataloging-in-Publication Data
A catalog record for this book is available from the Library of Congress

British Library Cataloguing-in-Publication Data
A catalogue record for this book is available from the British Library

ISBN: 978-0-12-819152-1

Publisher: Candice Janco
Acquisitions Editor: Amy Shapiro
Editorial Project Manager: Ruby Smith
Production Project Manager: Bharatwaj Varatharajan
Cover Designer: Mark Rogers

Typeset by STRAIVE, India

Dedication

To the GNSS community and its amazing power to transform and enhance our daily life and future dreams!

Contents

Preface

The book starts with the introduction of the GNSS and their components in Chapter 1 and the basics of signals and positioning in Chapter 2. Then it explains the basics of the atmosphere starting from top to bottom layers, i.e., from ionosphere in Chapter 3 to the troposphere in Chapter 4. Chapter 5 is dedicated to the numerical weather prediction models. The GNSS tropospheric monitoring is separated for application in numerical weather prediction and nowcasting. The methodologies behind profiling the atmosphere using GNSS signals are presented in Chapter 6. Chapters 7 and 8 focus on the application of GNSS for monitoring the climate and soil/vegetation/snow using direct and reflected GNSS signals. In Chapter 9 we come back to GNSS processing and introduce the latest developments for using atmospheric data to provide precise real-time GNSS products, which has the potential to open a new page in GNSS applications.

Foreword

GNSS monitoring of the atmosphere is an interdisciplinary topic: a collaboration between geodetic and atmospheric communities. An interdisciplinary topic requires sufficient basic knowledge about both GNSS and the atmosphere. Basic errors and/or misunderstanding of the atmosphere and its structure by PhD students and sometimes experienced researchers are seen in the scientific papers. On the other hand, the geodetic terminology and methodologies are often unclear to the meteorological community. Thus we took the time to develop a series of lectures introducing the fundamental concepts of atmospheric physics, observations, and modelling, along with GNSS observation techniques, tuned to both these communities. Those lectures are the basis of this book. Wc have been fortunate to do PhD studies at the time when GNSS monitoring of the atmosphere and hydrosphere were born and were still experimental techniques. It is shocking to a newcomer meteorologist in this GNSS community to know how much knowledge do geodesists have of the atmosphere and hydrosphere. To be able to tap in this knowledge we further developed the lectures into this handbook to enable the more complete study of the GNSS basics and processing. In short, we feel privileged to share our knowledge and experience with the new generation of PhD students. Even in our wildest dreams we did not think that GNSS will be a standard meteorological observing system today and it was only 20 years ago... Amazing isn't it!

Enjoy reading!

Guergana and Tzvetan

Acknowledgments

We express our gratitude to the colleagues, who helped us with the creation of this work. Specifically, Kyriakos Balidakis (GFZ Potsdam, Germany) and Tomas Hadas (WUELS, Poland), who provided illustrations for the book.

We are very grateful to our GNSS mentors, who introduced us to this topic and encouraged us to dig deeper into it. Sometimes literally.

CHAPTER 1

Global Navigation Satellite System (GNSS—GPS, GLONASS, Galileo)

Introduction

The global navigation satellite system (GNSS) is a generic term to describe the multitude of positioning systems, sharing similar purpose and capabilities, namely, to provide all-weather high-precision navigation service with global coverage. Currently, there are four GNSS systems, namely, the US-developed Global Positioning System (GPS), the Russian GLONASS, the European Galileo, and the Chinese BeiDou. Each individual GNSS consists of three segments: (1) space segment (the satellites in orbit), (2) ground control segment, and (3) ground- or space-based user segment.

All the GNSS satellites have similar technical parameters as presented in Table 1.1.

There are several reasons why the systems resemble each other.

The orbital height of 18,000–24,000 km (Fig. 1.1) is chosen, so that each satellite can cover the whole visible surface of the Earth from a relatively small solid angle.

The satellite orbit needs to be close enough to the Earth's surface to avoid large signal delays from long signal paths.

Fewer satellites in medium Earth orbit (MEO) are required in order to have enough satellites visible from any point on the Earth's surface when compared to low Earth orbit (LEO).

The minimum number of satellites required in each system for full operational capabilities all around the world is 24. For precise positioning, a minimum of four satellites are necessary.

The navigation signal frequencies in the L-band, between 1 and 2 GHz, penetrate the Earth's ionosphere and troposphere without substantial attenuation.

Shorter waves would attenuate in the lower atmosphere, while longer wavelengths would not penetrate the ionosphere at shallow angles.

Multiple frequencies for positioning are required to mitigate the influence of the ionosphere on the positioning accuracy. With more than one

Global Navigation Satellite System Monitoring of the Atmosphere
https://doi.org/10.1016/B978-0-12-819152-1.00007-6

Table 1.1 GNSS comparison.

	GPS	GLONASS	Galileo	BeiDou
Nominal number of satellites	24	24	27	27 in MEO 3 in IGSO 5 in GEO
Orbits	6 elliptical	3 circular	3 circular	3 MEO 3 IGSO 1 GEO
Orbit height	20,180 km	19,140 km	23.222 km	21,155 km MEO 35,786 km IGS and GEO
Orbit inclination	55°	64°8′	56°	55°30′
Orbital repetition	1 day	8 days	10 days	7 days
Main frequencies	L1: 1.575 GHz L2: 1.228 GHz L5: 1.176 GHz	G1: 1.602 GHz G2: 1.246 GHz G3: 1.205 GHz	E1: 1.575 GHz E5a: 1.176 GHz E5b: 1.192 GHz E6: 1.279 GHz	B1: 1.561 GHz B1C: 1.575 GHz B2a: 1.176 GHz B3: 1.269 GHz

Space segment parameters of individual GNSS system. All GNSSs use medium Earth orbit (MEO) exclusively, except BeiDou, which also uses inclined geosynchronous orbit (IGSO), as well as satellites in geostationary orbit (GEO).

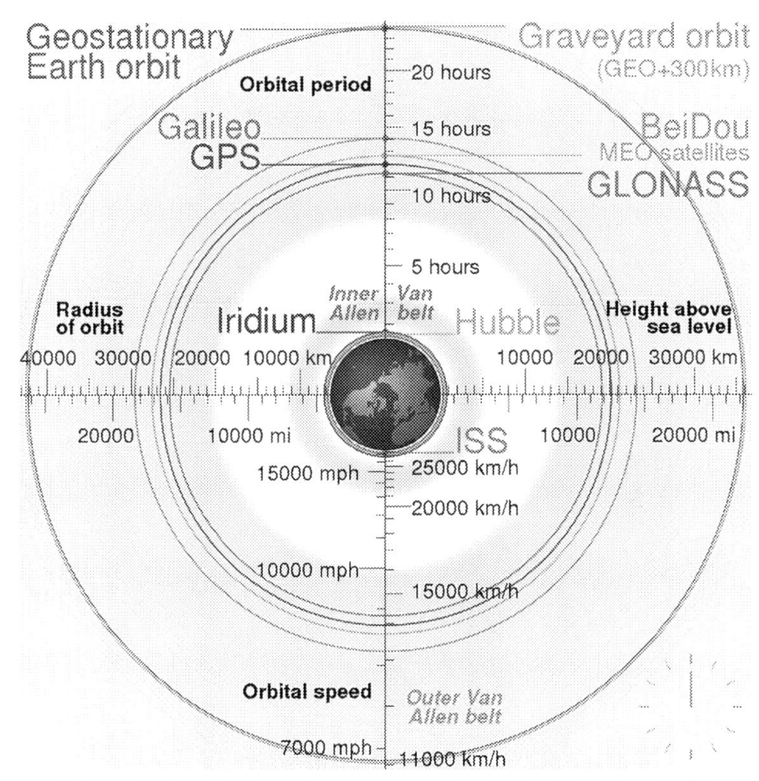

Fig. 1.1 Orbits of GNSS. Schematic representation of Earth's orbits. GNSS satellites occupy medium Earth orbits (MEO) with orbital heights between 18,000 and 24,000 km and orbital periods 10–14 h. *(Comparison satellite navigation orbits (n.d.). Retrieved September 27, 2020, from https://upload.wikimedia.org/wikipedia/commons/thumb/b/b4/Comparison_satellite_navigation_orbits.svg/1024px-Comparison_satellite_navigation_orbits.svg.png.)*

frequency for positioning the ionospheric delays of the signals can be calculated and accounted for.

All GNSS satellites are equipped with atomic clocks for high-precision timekeeping with stability within 30 ns (Teunissen & Montenbruck, 2017).

The satellites in orbit are tracked by the ground control segments of each GNSS. The control segments consist of several ground stations, spread around the globe, which monitor each individual satellite. The control centers of each GNSS can steer the satellites, switch them in different modes, or even transfer them between positions in the same orbit or in different orbits. All these commands are sent through the control segment via S-band signals to the satellites.

The third segment of the GNSS is the ground- or space-based user segment. It consists of all GNSS receivers used in reference ground-based positioning stations and mobile devices from mobile phones to satellites in orbit. Currently, there are more than 30,000 reference ground-based stations worldwide, installed for enhancing the GNSS capabilities and accuracy, as well as environmental research.

The complete description of the structure, engineering background, and operation of the GNSS is beyond the scope of this book. A more detailed description of the system's architecture and engineering principles can be found in the "Handbook of the Global Navigation Satellite Systems" and in "Global Positioning System: Theory and Practice" (Hofmann-Wellenhof, Lichtenegger, & Collins, 2012; Teunissen & Montenbruck, 2017).

Space segments of the GNSS

Global Positioning System (GPS)

Historically, the first GNSS was the GPS, developed by the US Department of Defense in the 1970s. GPS was not only the first GNSS to be designed, but also the first system to become fully operational and accessible to civilian users. This system dominated the satellite positioning for over three decades and has been the basis for multiple applications and observation techniques, which were later adapted to the other GNSS systems (see Table 1.2).

The GPS space segment comprises more than 24 satellites in six elliptical MEO at 55° inclination with respect to the Earth's equator (as described in Table 1.1). The satellites transmit at three frequencies: (1) L1 at 1575.42 MHz, (2) L2 at 1227.60 MHz, and (3) L5 at 1176.45 MHz. The reason to use more than one frequency for positioning is described in Chapter 3. The satellites from three generations are deployed. The first generation is named Block I and has 10 satellites.

Table 1.2 Publications numbers.

	GNSS	GPS	GLONASS	Galileo	BeiDou
Web of Science	13,566	73,181	1,823	7,667	4 238
Google Scholar	429,000	3,590,000	88,900	72,500	37 000
Science Direct	6,016	114,412	1,655	17,429	16 192

Number of publications with GNSS key words according to different academic search engines. GPS dominates the number of research articles found in all major databases.

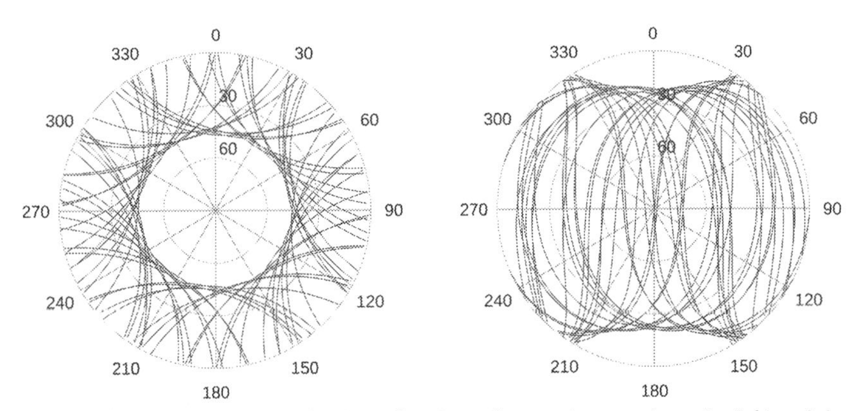

Fig. 1.2 Skyplots of the GPS. Skyplots of GPS satellite tracks over the pole *(left)* and the equator *(right)*. *(Courtesy of Kyriakos Balidakis.)*

The second generation has five iterative designs Block II, IIA, IIR, IIRM, and IIF and is the backbone of the system with 64 satellites. Throughout the lifespan of the second generation several improvements were introduced, including the new signals like the civilian L2C (with Block IIR) and L5 (with Block IIF). The first satellite from the third generation was launched in 2018 and is in operational service from 2020. The third-generation GPS satellites have an additional signal—the civilian L1C (https://www.gps.gov/).

The satellite constellation is designed to allow full Earth coverage with a minimum of six satellites visible at any time from any point on the Earth. The satellites have orbital period of 11 h 58 m and orbital repetition of 1 sidereal day. This means that the satellites rise and set in the same direction each day relative to each point on the Earth's surface. Fig. 1.2 presents the skyplots of the satellites visible at the pole (left) and the equator (right). At latitudes above 55° the positioning accuracy is reduced, since no GPS satellite is visible close to nadir (see satellite ground tracks in Fig. 1.3).

GLObal NAvigation Satellite System (GLONASS)

The GLONASS system was initiated in the Soviet Union as a strictly military project and the system reached limited operational status in the early 1990s but by 2001 only six satellites were operational in orbit. In the early 2000s, the GLONASS program was revived with a new, second generation of satellites (GLONASS-M) and the ambition to complete the constellation.

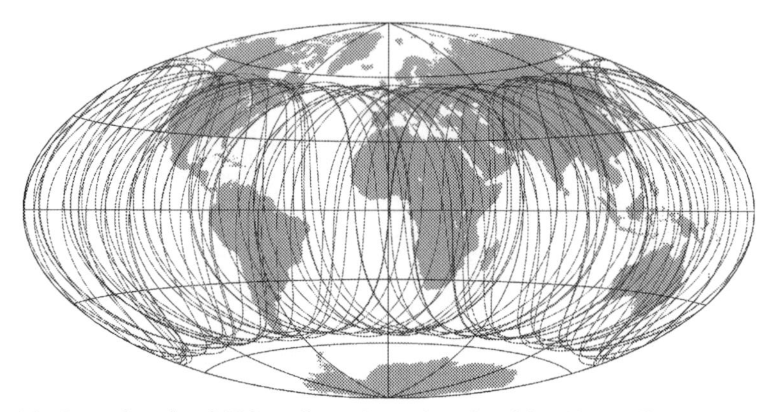

Fig. 1.3 Ground tracks of GPS satellites. Ground tracks of the GPS satellites. *(Courtesy of Kyriakos Balidakis.)*

The GLONASS was announced to be fully operational in 2015 (http://en.roscosmos.ru/).

The space segment of the system consists of more than 24 satellites in three circular orbits, inclined at 64° 8′ with respect to the equator. The higher inclination of the orbits secures better coverage and position accuracy in Earth's polar regions (Fig. 1.4). The satellites of GLONASS transmit at different frequencies for each satellite in a wide range around the main frequencies. This approach of identifying the satellites in orbit is known

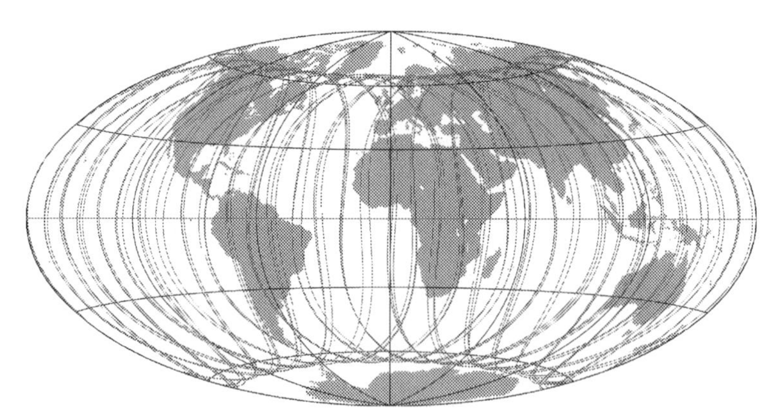

Fig. 1.4 Ground tracks of GLONASS. Ground tracks of the GLONASS satellites. *(Courtesy of Kyriakos Balidakis.)*

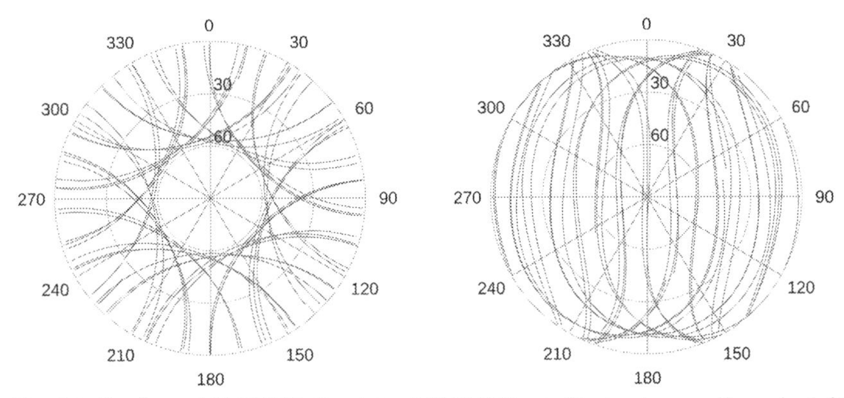

Fig. 1.5 Skyplots of GLONASS. Skyplots of GLONASS satellite tracks over the pole *(left)* and the equator *(right)*. *(Courtesy of Kyriakos Balidakis.)*

as frequency division multiple access (FDMA). This approach of satellite identification by the user segment was used for the first, second, and third (GLONASS-K) generation of satellites. The fourth generation of satellite (GLONASS-V, launch planned for 2025) is expected to use code division multiple access (CDMA), similar to GPS, where satellite identification is incorporated into the navigational message. The GLONASS satellites have an orbital period of 11 h 15 m. Unlike GPS, the orbital repetition of the constellation is 8 days. Fig. 1.5 shows the skyplots of GLONASS satellites over GNSS stations in polar (left) and equatorial (right) latitudes. The higher inclination of GLONASS orbits provides a smaller gap in the satellite coverage from the polar regions, compared to the other GNSS (https://www.glonass-iac.ru/en/).

European GNSS Galileo

The European GNSS Galileo was conceived in the 1990s as a joint effort by the member states of the European Union. Unlike GPS, GLONASS, or BeiDou, the Galileo is not intended as a military project and is aimed at civilian applications. The system has a predefined set of applications, including civilian navigation Open Service (OS), High Accuracy Service (HAS), Search and Rescue (SaR), and Public Regulated Service (PRS) for highly sensitive government-operated infrastructure uses. Unlike the other GNSS systems, the satellites of Galileo are equipped with hydrogen maser atomic

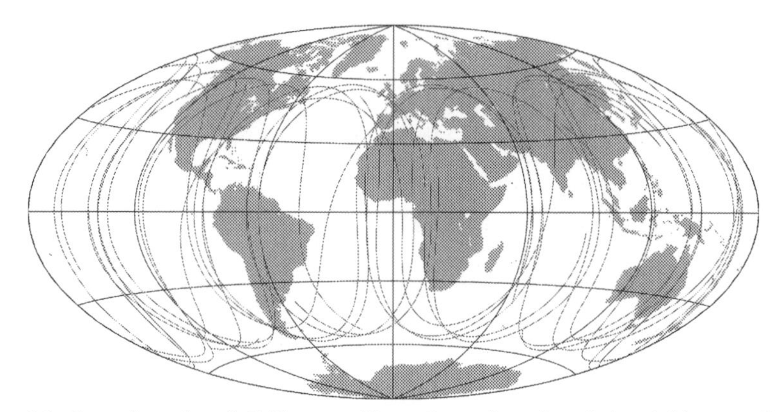

Fig. 1.6 Ground tracks of Galileo satellites. Ground tracks of the Galileo satellites. *(Courtesy of Kyriakos Balidakis.)*

clocks, which have much higher stability than the rubidium atomic clocks used in the other GNSS satellites. The first two satellites of the constellation were launched in 2011 with expected full constellation achieved in 2018. As of 2020, all the 26 launched satellites belong to the first generation with the second generation of satellites expected earliest in 2025 (http://www.esa.int/esaNA/galileo.html) (see ground tracks of the launched Galileo satellites in Fig. 1.6). The satellites were launched into three orbital planes with 56° inclination and orbital period of 14 h (see Table 1.1), providing coverage of the polar regions similar to GPS (see Fig. 1.7).

BeiDou (BDS) GNSS

The BeiDou (or BDS) GNSS system has been developed and operated by the Chinese National Space Agency (CNSA) since the 1990s. Three generations of the system were developed. BeiDou-1 is the first generation of the system, deployed as a regional satellite navigation system, consisting of three geostationary satellites (GEO, Fig. 1.8). The system was expanded through the BeiDou-2 to include satellites in MEO and inclined geosynchronous orbits (IGSO). These new satellites allowed to extend the regional coverage to global. With the addition of satellites from the BeiDou-3 generation, the system became fully operational in 2018 and is providing global coverage with a complete constellation since 2020. The BeiDou system uses GEO, IGSO, and MEO satellites, which makes it unique among the other GNSS

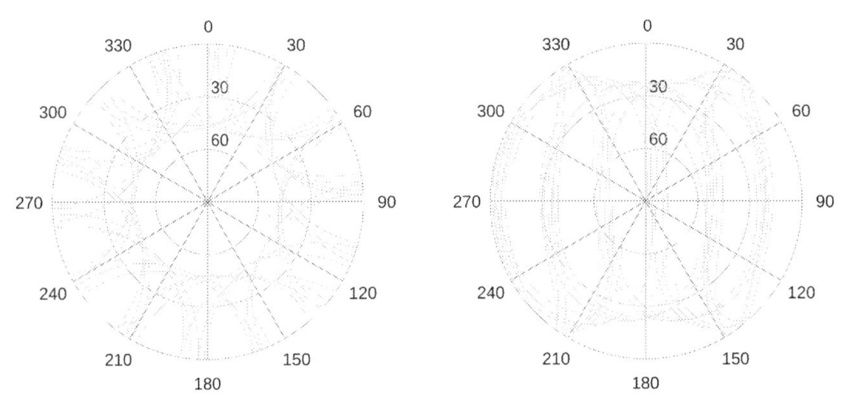

Fig. 1.7 Skyplots of Galileo satellites. Skyplots of Galileo satellite tracks over the pole *(left)* and the equator *(right)*. *(Courtesy of Kyriakos Balidakis.)*

(see ground tracks of BeiDou satellites in Fig. 1.8). The GEO and IGSO satellites are exclusively situated in the Eastern Hemisphere and intended to increase the number of available satellites over the territory of China and the Asian region (http://en.beidou.gov.cn/).

Ground-based reference GNSS networks

The user segment of the GNSS is one of the largest industries in the world economy. According to a 2015 estimate, GPS alone accounted for \$56

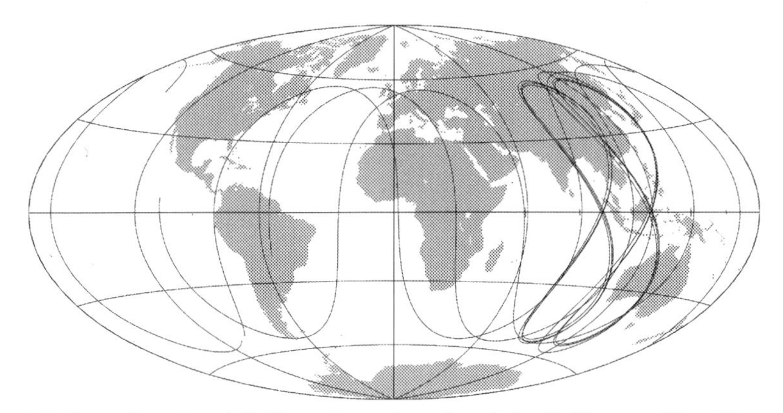

Fig. 1.8 Ground tracks of BeiDou. Ground tracks of the BeiDou satellites. Note the ground tracks of the IGSO satellites over China, Australia, and the Indian ocean. *(Courtesy of Kyriakos Balidakis.)*

billion of benefits to the US economy in 2013 (https://www.gpsworld.com/the-economic-benefits-of-gps/). The GNSS user segment consists of: (1) ground-based reference GNSS stations/networks and (2) mass market devices. The reference GNSS stations provide high-precision corrections for the mass market real-time users (mobile platforms) and, in addition, are used for scientific applications like monitoring of: ionosphere (Chapter 3), troposphere and climate (Chapters 4–7), and soils/vegetation/snow cover (Chapter 8). The International GNSS Service (IGS, http://www.igs.org) network consists of more than 500 ground-based reference stations (Fig. 1.9), distributed all around the world. The network members provide not only data from the stations, but also data about satellite orbits and clocks. These products are used to provide differential (DGNSS) corrections or to foster precise point positioning (PPP) solutions for stations and networks worldwide, increasing the accuracy of the positioning from tens of meters to tens of centimeters. A more detailed explanation of these processing techniques is given in Chapter 2.

Regional GNSS networks, such as the European Reference (EUREF) Permanent Network (EPN), provide similar services but on a regional scale. They are used for the following applications:

For providing data for the satellite orbits and clocks.

For determining ionospheric delays for monitoring the state of the ionosphere (GNSS-I described in Chapter 3).

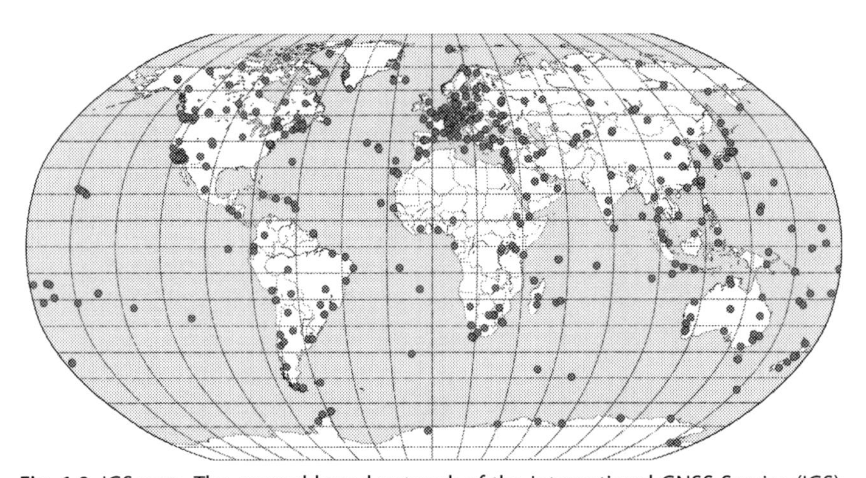

Fig. 1.9 IGS map. The ground-based network of the International GNSS Service (IGS).

For determining tropospheric delays for monitoring the water vapor distribution in the lower atmosphere through the GNSS meteorology (GNSS-M) method (see Chapter 4).

For monitoring snow height, soil moisture, and vegetation using GNSS reflected signals (GNSS-R described in Chapter 8).

For monitoring the Earth's crust movement.

For observing volcanic eruptions.

The European Meteorological Network (EUMETNET) has developed its own GNSS network for water vapor monitoring under the EUMETNET EIG GNSS water vapor program (EGVAP, http:/egvap.dmi.dk). Unlike the IGS and EPN, the EGVAP network is fully focused on GNSS-M applications. It provides data to national weather services for weather prediction and modeling, as well as for climate monitoring applications (see Chapter 5). The stations delivering data in near real-time mode (90 min after observation) to the EUMETNET are presented in Fig. 1.10.

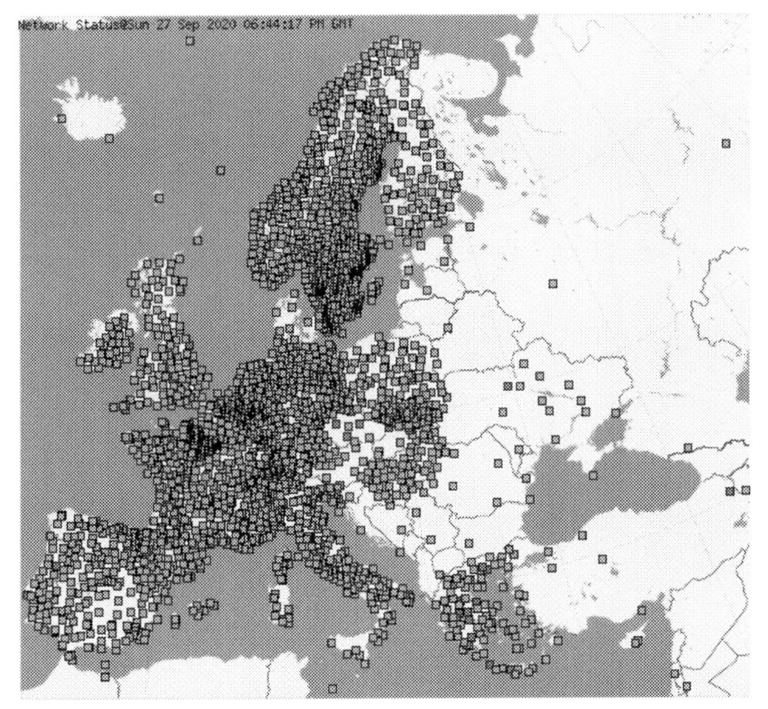

Fig. 1.10 EGVAP. The ground-based GNSS network of the European Meteorological Network EUMETNET, maintained throughout the EGVAP project. *(From http://projects.knmi.nl/egvap/validation/europe.html.)*

Positioning with GNSS

The process of position determination with GNSS for simplicity can be broken into three simple steps. Firstly, the GNSS satellites send signals to the Earth. The signal from each satellite carries a navigation message, containing the satellite identification, orbital position, and the time of transmission. Secondly, the receiver decodes the navigation messages from the satellites and calculates the signal travel time. In the third step, the receiver calculates the distance traveled by the signals from each satellite by multiplying the travel time with the speed of light.

The distance from one satellite creates a sphere around the satellite on which the position of the receiver lies (see Fig. 1.11). Incorporation of the distance from a second satellite confines the position of the receiver to the surface cross section of the two spheres from each of the satellites. By adding a third satellite, the position of the receiver is further specified

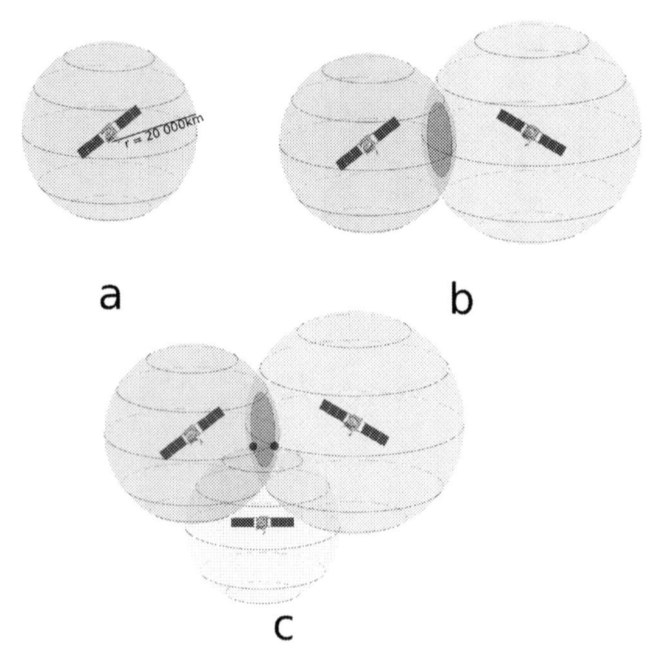

Fig. 1.11 Triangulation. Principle of triangulation using GNSS satellites. The distance from only one satellite (a) provides location on the surface of a sphere with the satellite in the center. When two satellites are used (b), the location is on the circle intersection of two spheres, centered around the satellites. When three satellites are available (c), the location is further constrained to two points on the circle from case b. By adding more satellites, more constraints to the location of the observer are implemented.

to two points on the circle. By the addition of distances to more and more satellites only one point remains, to which all distances converge. This positioning explanation, presented in Fig. 1.11, is extremely simplified, but represents the concept of triangulation. A more precise procedure for position determination through the navigation equations is presented in Chapter 2.

It is to be noted that the GNSS is designed primarily for positioning applications. The system is not intended as a provider of environmental information by design. Every application for environment observations, mentioned in this book, is a by-product of the availability of very precise and stable signals, penetrating through the Earth's atmosphere, providing positioning information with very high accuracy on a global scale.

References

Hofmann-Wellenhof, B., Lichtenegger, H., & Collins, J. (2012). *Global positioning system: Theory and practice*. Springer Science & Business Media.

Teunissen, P., & Montenbruck, O. (2017). *Springer handbook of global navigation satellite systems*. Springer.

CHAPTER 2

GNSS signals and basics of positioning

GNSS signal characteristics

GNSS uses L-band signals with frequency in the range 1–2 GHz and wavelength in the range 15–30 cm. More precisely, the GNSS systems use signals in two narrow bands: (a) a low L-band from 1.1 to 1.3 GHz and (b) an upper L-band from 1.55 to 1.65 GHz (see Fig. 2.1).

GNSS signals are electromagnetic waves and thus their interaction is described using Maxwell's dynamic theory of the electromagnetic field. Two distinguishable fields—electric (E) and magnetic (B) fields—are used to describe electromagnetic phenomena in uncharged medium using the following equations:

$$
\begin{aligned}
\nabla \cdot E &= 0 \\
\nabla \cdot B &= 0 \\
\nabla \times E &= -\frac{\delta B}{\delta t} \\
\nabla \times B &= \mu_0 \epsilon_0 \frac{\delta E}{\delta t}
\end{aligned}
\qquad (2.1)
$$

where $\nabla \cdot$ is the divergence operator, $\nabla \times$ is the curl operator, $\frac{\delta}{\delta}$ is a partial derivative, μ_0 is the absolute permeability, and ε_0 is the absolute dielectric permittivity of vacuum. Several basic principles of electromagnetic waves propagation can be extracted from Eq. (2.1), namely:

- electric and magnetic fields of an electromagnetic wave propagate in the same direction,
- electric and magnetic field oscillations are orthogonal (perpendicular) to each other,
- both electric and magnetic fields are orthogonal (perpendicular) to the propagation direction, and.
- both electric and magnetic fields maintain their direction of propagation—have no divergence.

Global Navigation Satellite System Monitoring of the Atmosphere
https://doi.org/10.1016/B978-0-12-819152-1.00004-0

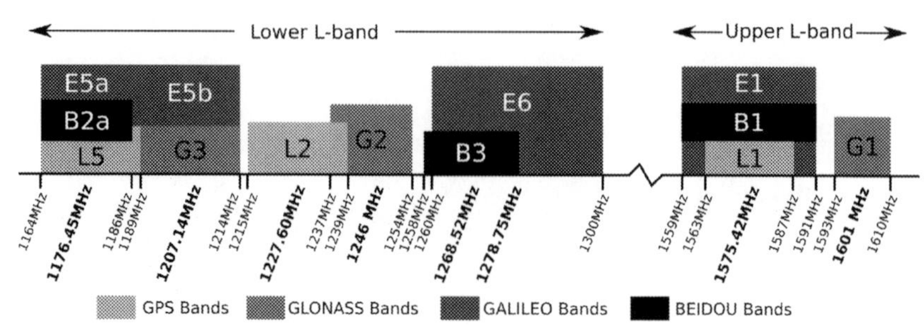

Fig. 2.1 GNSS frequency bands. L-band frequencies used by the GNSS-GPS are indicated in lightest gray, GLONASS in light gray, Gallileo in dark gray, and BeiDou in black. All GNSS have a single-frequency band in the upper L-band and multiple frequencies in the lower L-band.

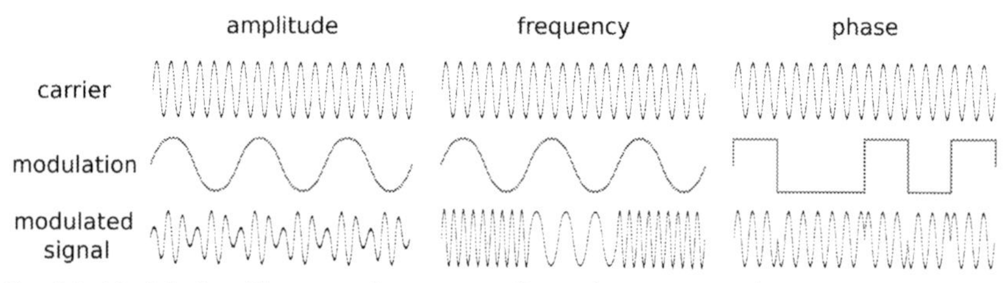

Fig. 2.2 Modulation. There are three ways of introducing a signal into a carrier wave—amplitude, frequency, and phase modulation. GNSS uses phase modulation to transmit binary-coded information from the satellites to the receiver.

One of the main properties of electromagnetic waves is their polarization. The polarization stands for the direction of the electric field plane of the electromagnetic wave. When the electric field vector over time oscillates on the same plane, the wave has linear polarization. When the electric field vector changes its direction over time following a screw-like pattern, the wave is elliptically or circularly polarized.

In the case of GNSS signals, the electromagnetic waves carrying the signal are right-hand (clockwise) circular polarized (RHCP).

In order to carry information, the electromagnetic waves need to be modulated. In the case of GNSS signals, a binary phase modulation is introduced into the signal. There are three distinct modulation types, which can be used to fuse information-carrying signal onto carrier radio frequencies (Fig. 2.2). Phase modulation is typically used for carrying binary data such as the GNSS navigation messages.

The GPS signals can be used as an example for all GNSS, since they employ similar design and capabilities. Each satellite transmits at several

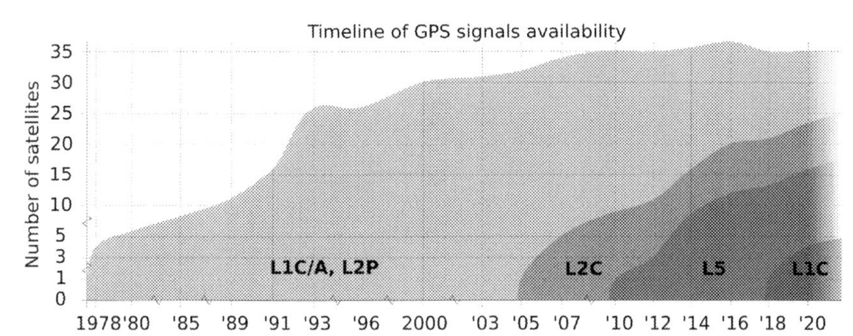

Fig. 2.3 GPS signal history. Historical availability of GPS civilian signals. *(From Simeonov, T. (2021). Derivation and analysis of atmospheric water vapor and soil moisture from ground-based GNSS stations. In Technische Universität Berlin. Technische Universität Berlin.)*

frequencies; in case of GPS, these are L1, L2, L3, L4, and L5. Each frequency carries several codes of information (see Fig. 2.3). Of these frequencies only L1, L2, and L5 carry civilian codes.

The most widely used civilian GPS code, the L1 Coarse Acquisition (L1C/A), uses Gold codes (Gold, 1967), modulated at 1.023 MHz, carrying 1023 chips each millisecond. With the development of the system a new code is being employed, the L1C, using the same modulation frequency, but carrying 10,230 chips, which provides a 10 times larger data bandwidth. The codes are identified by the receivers through small offsets in the main carrier frequency (see Fig. 2.4). Apart from the civilian L1C/A, L1C, L2C, and L5, a number of codes are restricted to military use only. They use similar principles of coding, but the messages can only be read by specialized receivers.

Each GNSS satellite message contains the following data:

- satellite identifier (PRN),
- ephemeris parameters of the satellite orbit,
- time parameters of the satellite onboard clocks,
- clock corrections,
- satellite health information,
- ionospheric parameters for single-frequency receivers,
- almanac, used for initial coarse acquisition of all available satellites.

While the ephemeris and time parameters are transmitted in every frame of the message, the ionospheric parameters and the almanac are transmitted in parts, thereby enabling the receiving of the full navigation message every 12.5 min for the L1C/A (Teunissen & Montenbruck, 2017).

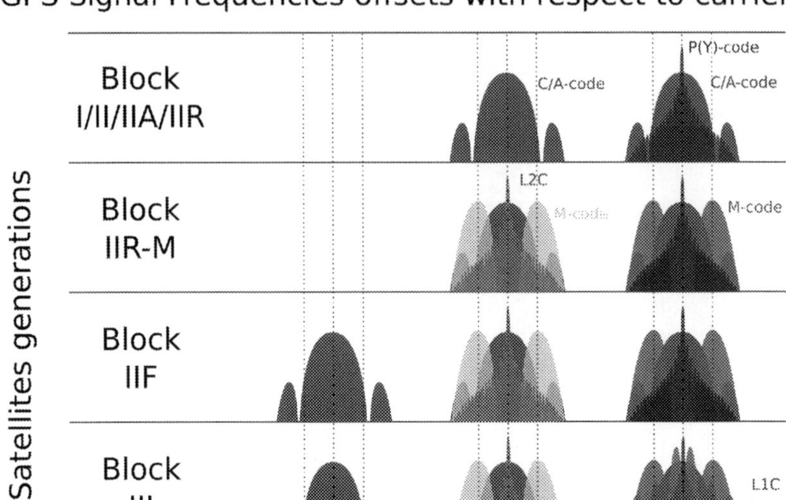

Fig. 2.4 GPS signal map. Structure of the GPS signal codes at L1, L2, and L5 frequencies. *(Recreated from Hegarty, C. J., & Chatre, E. (2008). Evolution of the global navigation satellite system (GNSS). Proceedings of the IEEE, 96(12), 1902–1917.)*

GNSS observation equations

The pseudorange measurements represent the apparent signal travel time between the GNSS satellite and the receiver. The receiver generates a replica of the transmitted satellite code and aligns it with the received signal. The time shift between the two codes is the apparent transit time of the signal (see Fig. 2.5). It is then combined with additional information from the satellite's navigation data to obtain the actual travel time from the satellite to the receiver. The travel time is then multiplied with the speed of light to obtain the pseudorange between the satellite and the receiver. These measurements differ from the actual distance, since the signal is subject to delays and the receiver and satellite clock offsets are unknown.

The pseudorange equation describing the distance between the satellite and the receiver takes the following form:

$$p_r^s(t) = \rho_r^s(t) + \xi_r^s(t) + c(d_r + d^s) + c(dt_r + dt^s + \delta t^{rel}(t)) + I_r^s(t) + T_r^s(t) + \epsilon_r^s(t) \tag{2.2}$$

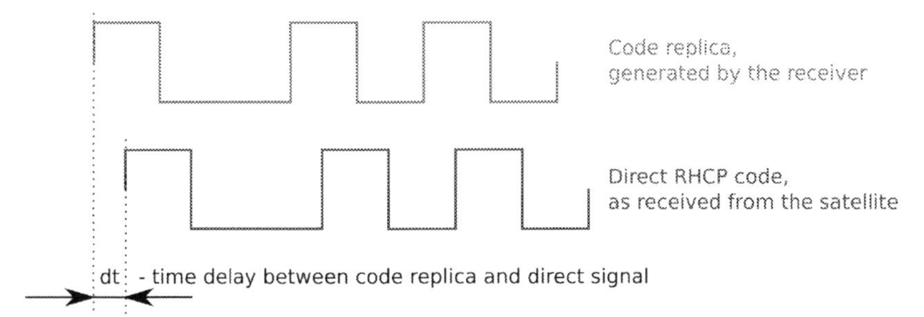

Fig. 2.5 Code replica. The receiver creates a replica of the code received from the satellite and measures the time difference between the direct signal and the replica to in turn measure the range to the satellite. *(From Simeonov, T. (2021). Derivation and analysis of atmospheric water vapour and soil moisture from ground-based GNSS stations. In Technische Universität Berlin. Technische Universität Berlin.)*

where $p_r^s(t)$ is the pseudorange, $p_r^s(t)$ is the actual distance between the satellite and the receiver, $\xi_r^s(t)$ is the correction of the phase-center offsets of the transmitting and the receiving antennae, d_r and d^s are the receiver and satellite instrumental delays, dt_r and dt^s are the clock offsets, $\delta t^{rel}(t)$ are relativistic corrections, I and T are the ionospheric and tropospheric delays, and $\epsilon_r^s(t)$ are residuals, such as noise and multipath.

The receiver also records the carrier phase angle. It creates a replica of the carrier signal, aligns it with the observed messages from the satellite and then measures the phase shift between the two. Since the wave lengths of the GNSS signals are in the range between 15 and 30 cm, each full phase cycle of 360° indicates that the change in the distance between the satellite and the receiver is equal to the wavelength. The carrier-phase measurements are more precise than the pseudorange measurements, because they are relative to one another. However, the carrier-phase observation cannot be used to calculate the absolute distance between the receiver and the satellite, but only relative changes. The carrier-phase observation equation is given by

$$\begin{aligned}\phi_r^s &= \rho_r^s(t) + \xi_r^s(t) + c(\delta_r + \delta^s) + c\left(dt_r + dt^s + \delta t^{rel}(t)\right) - I_r^s(t) + T_r^s(t) \\ &\quad + \lambda(\omega(t) + N) + \epsilon_r^s(t)\end{aligned} \tag{2.3}$$

where λ is the wave length of the signal, ω is the relative angular rotation between the receiver and the transmitter antennae, and N is an integer number of cycles, called ambiguities. Different level of positioning accuracy is achieved through the treatment of the ambiguities: firstly real ($\mathbb{R}$) ambiguities are retrieved through least-square estimations, secondly the ambiguities are mapped into integers ($\mathbb{R} \rightarrow \mathbb{Z}$) and lastly the integer ambiguities are fixed and a second least-square adjustment is carried out for the final positioning.

Another observable by the receiver signal characteristic, is the Doppler shift of the received frequency. The Doppler shift is caused by the relative movement between the satellite and the receiver and can be expressed by the following equation:

$$D_r^s = \frac{1}{\lambda}\left(\frac{\vec{v^s}}{c} - \vec{e}\right) \cdot \left(\vec{v^s} - \vec{v_r}\right) + \left(df_r + df^s\right) + \frac{c}{\lambda}\delta f_{clk}^{rel} \tag{2.4}$$

where D is the observed Doppler shift, df_r and df^s are the frequency deviations of the receiver and the satellite, c is the speed of light, $\vec{v_r}$ and $\vec{v^s}$ are the relative movement of the receiver and the satellite, respectively, $\vec{e}$ is the unit light of sight between the satellite and the receiver, and δf_{clk}^{rel} are the clock-related relativistic effects (Parkinson, Enge, Axelrad, & Spilker Jr, 1996; Spilker Jr, Axelrad, Parkinson, & Enge, 1996).

Atmospheric refraction

The atmosphere is a medium with changing density, where density decreases with increase in altitude. The electromagnetic waves traveling through such medium with changing density experience a decrease in their speed, according to the optical density of the atmosphere. Following Snell's law, the optical density (also known as refractive index) of a medium is described by the speed of electromagnetic waves passing through it:

$$n_m = \frac{\text{Speed of light in vacuum}}{\text{Speed of light in the medium}} = \frac{c}{v} \tag{2.5}$$

Snell's law postulates that an electromagnetic wave, penetrating the border between two media with different optical densities, changes the direction of its propagation:

$$n_1 \sin \alpha_1 = n_2 \sin \alpha_2 \tag{2.6}$$

where α_1 is the propagation angle in the first medium, α_2 is the angle in the second medium, and n_1 and n_2 are the refractive indices of the media (see Fig. 2.6). Fermat's principle is the integrated form of Snell's law for a medium with gradually changing optical density:

$$S = \int_a^b n(s)ds \tag{2.7}$$

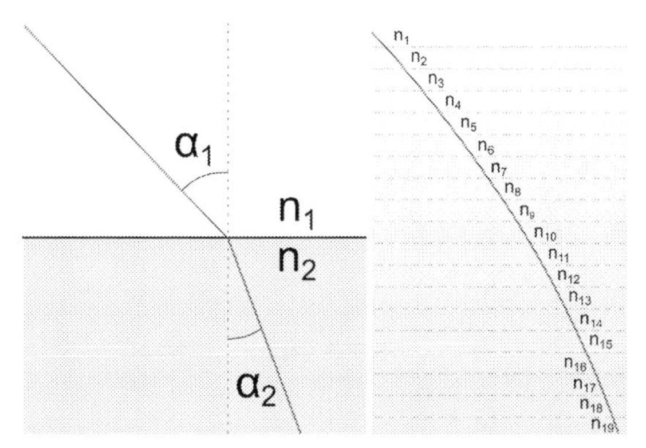

Fig. 2.6 Snell's law and Fermat's principle. Visualization of Snell's law and Fermat's principle. *(From Simeonov, T. (2021). Derivation and analysis of atmospheric water vapour and soil moisture from ground-based GNSS stations. In Technische Universität Berlin. Technische Universität Berlin.)*

where S is the optical path of the wave through the medium with changing density ($n(s)$), and a and b are the start and the end of this path, respectively. For the atmosphere this equation can be modified to the following:

$$S = \int_{h_0}^{h_{top}} n(h)dh \tag{2.8}$$

where h_{top} is the top of the atmosphere and h_0 is the Earth's surface.

The optical density of the atmosphere (n) depends on its pressure (p is the pressure, p_0 is the pressure at sea level), temperature (T is the temperature, $T_0 = 273.15\,\mathrm{K}$ is the melting point of water), and properties of the air molecules (N_a is the Avogadro's constant, V_0 is the molar volume of an ideal gas under standard conditions, α is the scalar atomic polarizability, and ϵ_0 is the dielectric permittivity of vacuum) and can be described by the following equation (Foelsche, 1999):

$$n = \frac{10^6 N_a T_0}{2\epsilon_0 V_0 p_0} \alpha \frac{p}{T} \tag{2.9}$$

Optical density is a measure of the ratio between the speed of propagation in vacuum and the speed of light in a certain medium. The difference between the time needed for the signal to travel in vacuum and the signal to travel in the medium is referred to as delay.

Mapping functions

GNSS receivers receive positioning messages from the satellites at elevation angles close to the horizon up to the zenith. Thus, the signals travel through longer or shorter slanted paths to the receiver, depending on the angle and the thickness of the atmosphere at the specific locations, so that each signal is delayed differently.

The mapping function is a projection of the tropospheric wet and dry delays to zenith (as seen in Fig. 2.7). The projection depends on the elevation angle (ϵ) of the satellite. The simplest mapping function can be derived as.

$$m(\epsilon) = \frac{1}{\sin(\epsilon)} \tag{2.10}$$

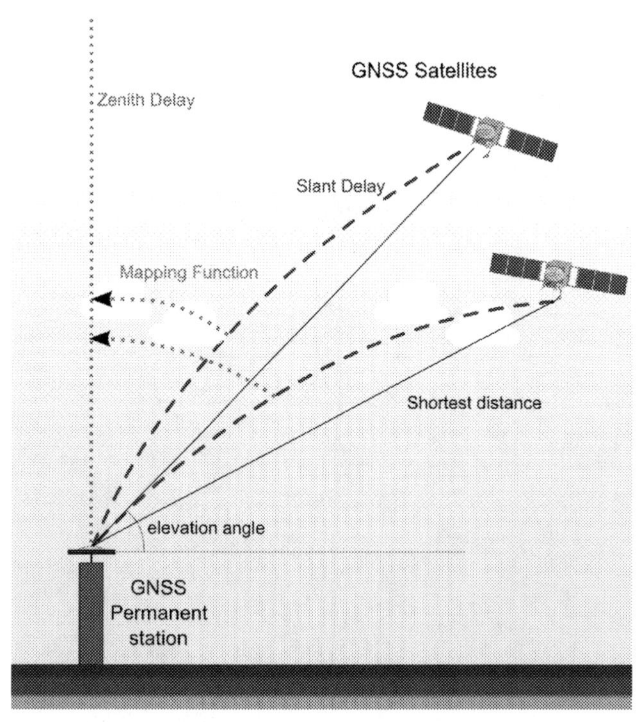

Fig. 2.7 Mapping function. Effect of the neutral atmosphere on the GNSS signals. The electromagnetic waves follow the optical path, defined by Fermat's principle. The tropospheric delays are then mapped to the zenith. *(From Simeonov, T. (2021). Derivation and analysis of atmospheric water vapour and soil moisture from ground-based GNSS stations. In Technische Universität Berlin. Technische Universität Berlin.)*

but this approximation is far from perfect. Marini (1972) developed a more complex and accurate mapping function:

$$m(\epsilon) = \cfrac{1}{\sin(\epsilon) + \cfrac{a}{\sin(\epsilon) + \cfrac{b}{\sin(\epsilon) + \ldots}}} \tag{2.11}$$

where a and b are coefficients, defined differently by different authors. Arther Niell (1996) proposed a set of mapping functions (Niell Mapping Function, NMF) as follows:

$$m(\epsilon) = \frac{\cfrac{1}{1 + \cfrac{a}{1 + \cfrac{b}{1 + c}}}}{\sin(\epsilon) + \cfrac{a}{\sin(\epsilon) + \cfrac{b}{\sin(\epsilon) + c}}} \tag{2.12}$$

The Global Mapping Function (GMF) is a variation of the NMF, where b and c are empirically derived values, while a has the following structure (Böhm, Niell, Tregoning, & Schuh, 2006):

$$a = a_0 + A\cos\left(\frac{doy - 28}{365} 2\pi\right), \tag{2.13}$$

where a_0 is a global grid of mean values, A is a global grid of amplitudes for both hydrostatic and wet coefficients, and doy is day of year. Crucially, a, b, and c coefficients of the GMF are determined using more sophisticated atmospheric information—data from the Global Pressure and Temperature (GPT) model. Thus, GMF and the similar Vienna Mapping Function (VMF) are not purely geometrical, but based on atmospheric properties (Kouba, 2008). The differences between GMF, VMF, and NMF can reach up to 10 mm vertically (Böhm et al., 2006).

Zenith tropospheric delay

The atmosphere is composed of different gases, each with its own optical density. Based on the optical density of the wet and dry constituents of the atmosphere, delays for dry (Z^{dry}) and wet (Z^{wet}) atmosphere can be postulated:

$$d_{trop}^{dry} = \int_{h_0}^{h_{top}} \left(n^{dry}(h) - 1\right) dh, \tag{2.14}$$

$$d_{trop}^{wet} = \int_{h_0}^{h_{top}} (n^{wet}(h) - 1)dh. \tag{2.15}$$

These factors reveal how the dry gases and water vapor differ from ideal gas. The full tropospheric delay, as defined in Eqs. (2.2), (2.3), is a product of dry and wet delays:

$$d_{trop} = d_{trop}^{wet} + d_{trop}^{dry}. \tag{2.16}$$

The accuracy of the used mapping functions is very important for the accuracy of the computed zenith tropospheric delay (ZTD):

$$ZTD = m^{wet}(\epsilon)d_{trop}^{wet} + m^{dry}(\epsilon)d_{trop}^{dry}, \tag{2.17}$$

where ${d_{trop}}^{dry}$ is the dry tropospheric delay in the direction of the satellite, ${d_{trop}}^{wet}$ is the wet tropospheric delay, $m^{wet}(\epsilon)$ and $m^{dry}(\epsilon)$ are wet and dry mapping functions, respectively, at the elevation angle ϵ (Teunissen & Montenbruck, 2017). The tropospheric delay mapped to the zenith direction is measured in units of length, taking values between 1.5 and 3 m. Since the ZTD accounts for both dry and wet atmospheric components, it is used for the determination of atmospheric water vapor, as described in detail in Chapter 4.

GNSS data processing strategy

There are two approaches for processing GNSS observables: precise point positioning (PPP) and differential processing (DGNSS). The DGNSS technique is the older approach of the two. DGNSS relies on the availability of many GNSS stations, which are processed together. From the direct pseudorange and carrier-phase observations at each station, position differences between the stations are estimated (Hatch, 1989), usually referred to as baseline coordinates. The pseudorange differences are used in standard DGNSS processing, while the more accurate carrier-phase differential observations enable real-time kinematic solutions (RTK). Satellite parameters, such as frequency deviations, clock offsets, and orbits can be calculated with higher precision when large networks of stations with long baselines are processed together in the DGNSS processing mode (Teunissen & Montenbruck, 2017).

Another approach in GNSS processing is the PPP strategy (Zumberge, Heflin, Jefferson, Watkins, & Webb, 1997). In contrast to DGNSS, PPP uses data only from the station of interest, as well as GNSS satellite orbits and clock products of DGNSS processing. PPP is preferable for individual stations, or small dense networks, since it uses preprocessed clocks. Since 2013, the International GNSS Service (IGS, Dow, Neilan, Weber, & Gendt, 2007) has been providing ultrafast or real-time precise satellite orbit and clock corrections in support of PPP processing (Ahmed et al., 2016; Dousa & Vaclavovic, 2014; Li et al., 2015). The PPP strategy has the advantage of being computationally much more efficient than DGNSS and hence can provide estimates for large networks of stations with high temporal resolution (every 5 min). This task can be achieved by the conventional DGNSS strategies only by using superior IT infrastructure.

References

Ahmed, F., Vaclavovic, P., Teferle, F. N., Douša, J., Bingley, R., & Laurichesse, D. (2016). Comparative analysis of real-time precise point positioning zenith total delay estimates. *GPS Solutions, 20*(2), 187–199.

Böhm, J., Niell, A., Tregoning, P., & Schuh, H. (2006). Global Mapping Function (GMF): A new empirical mapping function based on numerical weather model data. *Geophysical Research Letters, 33*(7). https://doi.org/10.1029/2005GL025546.

Dousa, J., & Vaclavovic, P. (2014). Real-time zenith tropospheric delays in support of numerical weather prediction applications. *Advances in Space Research, 53*(9), 1347–1358.

Dow, J., Neilan, R., Weber, R., & Gendt, G. (2007). Galileo and the IGS: Taking advantage of multiple GNSS constellations. *Advances in Space Research, 39*(10), 1545–1551.

Foelsche, U. (1999). *Tropospheric watervapor imaging by combination of spaceborne and ground-based GNSS sounding data.*

Gold, R. (1967). Optimal binary sequences for spread spectrum multiplexing (Corresp.). *IEEE Transactions on Information Theory, 13*(4), 619–621.

Hatch, R. R. (1989). *Method for precision dynamic differential positioning.* Google Patents.

Kouba, J. (2008). Implementation and testing of the gridded Vienna Mapping Function 1 (VMF1). *Journal of Geodesy, 82*(4–5), 193–205.

Li, X., Dick, G., Lu, C., Ge, M., Nilsson, T., Ning, T., et al. (2015). Multi-GNSS meteorology: Real-time retrieving of atmospheric water vapor from BeiDou, Galileo, GLONASS, and GPS observations. *IEEE Transactions on Geoscience and Remote Sensing, 53*(12), 6385–6393.

Marini, J. W. (1972). Correction of satellite tracking data for an arbitrary tropospheric profile. *Radio Science,* 7(2), 223–231.

Niell, A. (1996). Global mapping functions for the atmosphere delay at radio wavelengths. *Journal of Geophysical Research: Solid Earth, 101*(B2), 3227–3246.

Parkinson, B. W., Enge, P., Axelrad, P., & Spilker, J. J., Jr. (1996). *Global positioning system: Theory and applications, Volume II.* American Institute of Aeronautics and Astronautics.

Spilker, J. J., Jr., Axelrad, P., Parkinson, B. W., & Enge, P. (1996). *Global positioning system: Theory and applications, Volume I.* American Institute of Aeronautics and Astronautics.

Teunissen, P., & Montenbruck, O. (2017). *Springer handbook of global navigation satellite systems.* Springer.

Zumberge, J., Heflin, M., Jefferson, D., Watkins, M., & Webb, F. (1997). Precise point positioning for the efficient and robust analysis of GPS data from large networks. *Journal of Geophysical Research: Solid Earth, 102*(B3), 5005–5017.

CHAPTER 3

Surveying the ionosphere with GNSS (GNSS-I)

Space weather

Sun and solar wind: Magnetosphere

The Earth, as a part of the galaxy, is under the constant influence of the interplanetary environment and as a part of the solar system it is strongly influenced by the Sun. Solar radiation provides the energy (heat and light) necessary for the living organisms but parts of the solar spectrum and high-energy cosmic rays are deadly to living organisms. The Earth's magnetic field is a shield that protects life on Earth from the harmful effects of cosmic rays, solar radiation, and more. According to the NASA definition, "the term space weather generally refers to conditions on the Sun, in the solar wind, and within Earth's magnetosphere, ionosphere, and thermosphere that can influence the performance and reliability of space-borne and ground-based technological systems and can endanger human life or health" (NASA Space Weather, n.d.). The Sun is a large energy source that produces high-energy protons which the solar wind carries to the Earth. During strong solar flares, energized protons can reach the Earth in 30 min after the eruption peak and the Earth is showered with high-energy solar particles (mainly protons). When these protons enter the atmosphere above the polar regions, it causes increased ionization at altitudes below 100 km. Ionization at these low altitudes is particularly conducive to the absorption of high-frequency radio signals and can make high-frequency communications in the polar regions impossible. This event is known as a polar cap absorption event.

During the solar minimum, the Sun's magnetic field is similar to the Earth's magnetic field, i.e., like an ordinary rod-shaped magnet with closed lines near the equator and open lines of force near the poles. Objects that have two distinct poles are called dipoles. The intensity of the magnetic dipole field of the Sun is about 50 G, which is equivalent to the intensity of the magnetic field of a refrigerator magnet. In comparison to the Sun, the Earth's magnetic field is about 100 times weaker. Around the solar

https://doi.org/10.1016/B978-0-12-819152-1.00010-6

maximum, when the Sun reaches its maximum activity, many spots are observed on the visible solar disk. These spots are a source of large magnetic forces that move material along with them. These local magnetic field lines are often hundreds of times stronger than the surrounding dipole. The Sun's magnetic field is not found only around the Sun itself. The solar wind carries it through the solar system until it reaches the heliopause (NOAA Solar Wind, n.d.). The heliopause is where the solar wind stops when it collides with interstellar space. That is why we call the Sun's magnetic field an interplanetary magnetic field. The interplanetary magnetic field has a spiral shape called the Parker spiral (Fig. 3.1), because the Sun revolves around its axis with a rotation period of around 27 days.

The interplanetary magnetic field is a vector quantity with components along the three axes, two of which (Bx and By) are oriented parallel to the ecliptic plane. Components Bx and By are not important for auroral activity (NOAA Aurora, n.d.). The third component Bz is perpendicular to the ecliptic plane and is caused by waves and other disturbances in the solar wind. When the interplanetary magnetic field and the geomagnetic field lines are inversely oriented or not mutually perpendicular, the geomagnetic field lines fail to remain closed, resulting in the transfer of energy, mass, and velocity from the solar wind flow to the Earth's magnetosphere. The strongest connection, with the most dramatic effects, occurs when the Bz component is strongly inclined to the south.

Fig. 3.1 Sun heliospheric current sheet—Parker spiral. *(Parker Spiral. (n.d.). https://en.wikipedia.org/wiki/Heliospheric_current_sheet#/media/File:Heliospheric-current-sheet.gif.)*

Space weather and GNSS

The space weather impact on GNSS signal propagation results in signal bending when passing through the charged plasma, i.e., similar to the way a lens bends the path of light (NOAA Space Weather, n.d.).

Ionospheric effect on electromagnetic wave propagation

Vertical structure of the atmosphere: Ionosphere

Different criteria can be used to determine the vertical structure of the atmosphere. The most widely used criteria take into account the characteristic changes of temperature in a given layer. For example, the troposphere is defined as a layer starting from the Earth's surface and spreading to an average of 12km. The troposphere is characterized by a temperature decrease with altitude. The stratosphere extends between 14 and 50km and temperature increases with altitude, while in the mesosphere (50–100km) temperature decreases and in the thermosphere (100–1000km) it increases (as shown in Fig. 3.2 left). The presence or the absence of ionized gases is another criterion for determining the vertical structure of the atmosphere. According to this criterion, the atmosphere can be divided into neutral, from the Earth's surface to 50km, and ionized from 50 to 600km (as shown in Fig. 3.2 right). The ionosphere is the part of the high atmosphere and is composed of ionized gases. Ionization is the result of the free electrons obtained as a result of

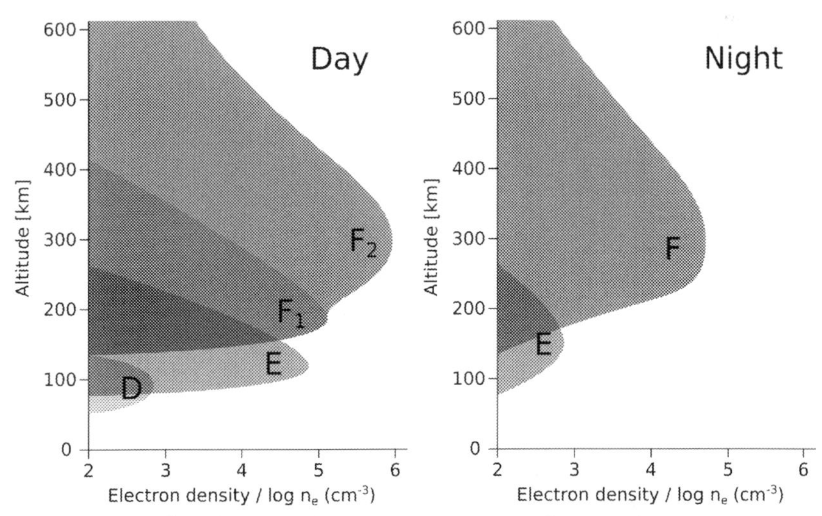

Fig. 3.2 Layers of Earth's ionosphere. *(Courtesy Tzvetan Simeonov.)*

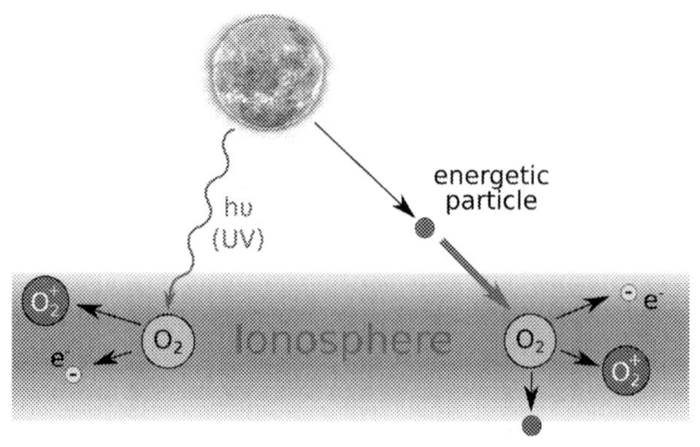

Fig. 3.3 Atmospheric ionization. *(Courtesy Tzvetan Simeonov.)*

the collision of atmospheric gases (nitrogen, oxygen) with high-energy particles or photons (Fig. 3.3). Ionization produces positively charged ions and free electrons, the content of which depends on solar activity. The free electron content has a maximum at an altitude between 200 and 400 km and is characterized by well-defined diurnal fluctuations. The ionosphere is composed of three main layers known as the D, E, and F layers (Fig. 3.2 right). The highest electron density is in the F layer. The ionization in the F layer is due to solar radiation during the day and due to cosmic rays during the night. During the day, the solar radiation (X-ray and UV) creates the D layer, and enhances the E layer as well as splits the F layer into two layers (Solar Flares Tracking, n.d.). The E layer reflects the radio signals back to the Earth's surface. This reflected signal is what Marconi used for the first long-distance radio communication in 1901 ("GNSS Ionosphere Explained," n.d.).

Ionospheric refraction of the GNSS signal

When the GNSS signal travels between the satellite and ground-based receiver, it is affected by the charged particles in the ionosphere. The distance traveled by the signal is increased, as described by the code navigation equation in Chapter 2. The phase of the signal is advanced (see the phase navigation equation in Chapter 2), which means that the impact of the ionosphere on the GNSS signals is frequency dependent. The refraction of the ionosphere is an important factor in the propagation of electromagnetic waves. As the GNSS code and carrier frequencies are different, the refractive index is defined for: (1) the carrier-phase refractive index (n_p, related to the carrier-phase pseudorange) and (2) the group refractive index (n_g, related to the code pseudorange). The relation between carrier and group refractive indices is given by

$$n_g = n_p + f\,\frac{dn_p}{df} \tag{3.1}$$

As a first-order approximation, the carrier-phase refractive index relation with the free electron content (n_e) or total electron content (TEC) is given by

$$n_p = 1 - \frac{40.3\,n_e}{f^2} \tag{3.2}$$

Here f is the frequency. The contribution of second-, third- and higher-order ionospheric terms is found to be 0.1% of the total (Hernández-Pajares et al., 2014). Substitution of Eq. (3.2) into Eq. (3.1) gives the following:

$$n_g = 1 + \frac{40.3\,n_e}{f^2} \tag{3.3}$$

From Eqs. (3.2), (3.3), it can be seen that the refraction index for the group and carrier phase are equal in magnitude but with different signs. The GNSS signal delay (S) caused due to its passing by the ionosphere is proportional to the total electron content along the path between the satellite and the receiver, i.e., slant total electron content (STEC) and is given by

$$S = \int_s^r n_e\,dl. \tag{3.4}$$

The STEC is commonly expressed in TEC units (TECU). 1 TECU is equal to 10^{16} e^-/m^2 and corresponds to a delay of about 16.2 cm for GPS L1 frequency.

The ionosphere range and phase error (ΔI_r^s) for a dual-frequency GNSS measurements is given by

$$\Delta I_r^s = 40.3\frac{f_1^2 - f_2^2}{f_1^2 f_2^2}\,TEC + \varepsilon \tag{3.5}$$

where ε is a residual noise term.

International reference ionosphere

The International Reference Ionosphere (IRI, n.d.) an empirical model for the terrestrial ionosphere was started in 1968 and evolved to international ISO standard in 2014 (Bilitza, 2018). By design, IRI is primarily based on observations from all available ground and space data sources and does not depend on the evolving theoretical understanding of ionospheric processes. IRI is the synthesis of all available and reliable ionospheric data, namely: (1) worldwide network of ionosondes, (2) the powerful incoherent scatter radars (Jicamarca, Arecibo, Millstone Hill, Malvern, St. Santin), (3) the topside sounders (ISIS and Alouette), and (4) in situ instruments on satellites and rockets. IRI provides monthly averages of (1) electron density and TEC, (2) electron temperature, (3) ion temperature, and (4) ion composition (O^+, H^+, He^+, N^+, NO^+, O^{+2}, cluster ions) in the altitude range 50–2000 km (Bilitza, 2018). IRI reliability depends on the spatiotemporal data coverage with good accuracy for data-rich northern mid-latitudes and not as good in data sparse polar and equatorial regions. The IRI model has four major inputs/drivers: (1) solar, magnetic, and ionospheric indices deduced from solar, magnetic, and ionospheric measurements, (2) equivalent solar or ionospheric indices obtained by adjusting the model to ionospheric measurements, (3) direct updates locally if measurements of characteristic parameters are available, and (4) assimilation of available data into the IRI background model (Bilitza, 2018).

Remote sensing of the ionosphere

Ionosonde

Stewart in 1878 first postulated the existence of a conductive layer in the upper atmosphere to explain variations observed in the Earth's magnetic field. In 1901, Marconi successfully transmitted a radio signal with frequency of 300 kHz across the Atlantic Ocean, between Cornwall, England, and Newfoundland, Canada. In 1902, Heaviside and Kennelly both independently suggested that Marconi's signal was propagated by reflection of signal from an ionized layer in the upper atmosphere. However, Appleton and Barnett were the first who proved the existence of the ionosphere in 1925, and were able to determine the radio reflection altitude. The unique relationship between the frequency of radio signal and the electron density, which can reflect the signal from the ionosphere is used for building the first instrument for ionospheric monitoring, called ionosonde. It works on the principle of a transmitter sending successive pulses of radio waves from 2 to 20 MHz into the ionosphere. These pulses are reflected at different layers and their reflections are registered by a ground-based receiver and analyzed by the control system. The result is displayed in the form of an ionogram (Ionogram, n.d.). Ionogram is a graph of reflection height (actually defined by the time between transmission and reception of a given pulse) vs the frequency. The ionograms are used for the calculation of the electron density profiles.

Global Ionosphere Radio Observatory (GIRO, n.d.) is an online data processing and dissemination center with an open data portal for global ionosonde observations. The ionosonde network (Fig. 3.4) provides data for the UN International Space Weather Initiative.

GNSS radio occlusion (GNSS-RO) missions

The extraction of information about the state of the atmosphere by processing a signal transmitted between two satellites is the basis of the radio occultation. The use of radio occultation in satellite planetary missions started in 1964 with a study of the atmosphere of Mars. In 1988, radio occultation was successfully applied to the GNSS signal for sounding the Earth's atmosphere. The first occultation experiment was a GPS/MET mission developed by the University Corporation for Atmospheric Research, Boulder, Colorado. To implement the method of radio occultations, satellites in low Earth orbit (LEO, 400 km see Fig. 1.1) are needed to intercept a radio signal passing through the Earth's atmosphere sent by a GNSS transmitter located on MEO satellite (20,000 km). The GNSS-RO receiver detects the change in the signal after passing through the different layers of the atmosphere. The main characteristics of GNSS-RO measurements are: high accuracy, high temporal and spatial resolution, and performing measurements in all weather conditions during the day and at night. The three GNSS-RO missions for ionospheric monitoring are presented here. Scientific Application Satellite-C (SAC-C) is an international satellite project for monitoring the structure and dynamics of the Earth's atmosphere, ionosphere, and geomagnetic field (SAC-C, n.d.). The satellite was launched into orbit on November 21, 2000. The orbit has an inclination of 98.2 degree with respect to the Earth's equator and is close to a polar orbit. SAC-C altitude is 702 km and the repetition period is 93–97 min. The orbit is synchronized with the Sun. The operational part of the work is performed by the NASA's Jet Propulsion Laboratory (JPL).

Another GNSS-RO satellite mission is the CHAllenging Minisatellite Payload (CHAMP)—a German satellite project for geophysical and atmospheric research led by the German Research Center for Geophysics (GFZ). It is equipped with the instruments onboard such as a magnetometer, a GNSS-RO receiver, an accelerometer, an ion concentration meter, and

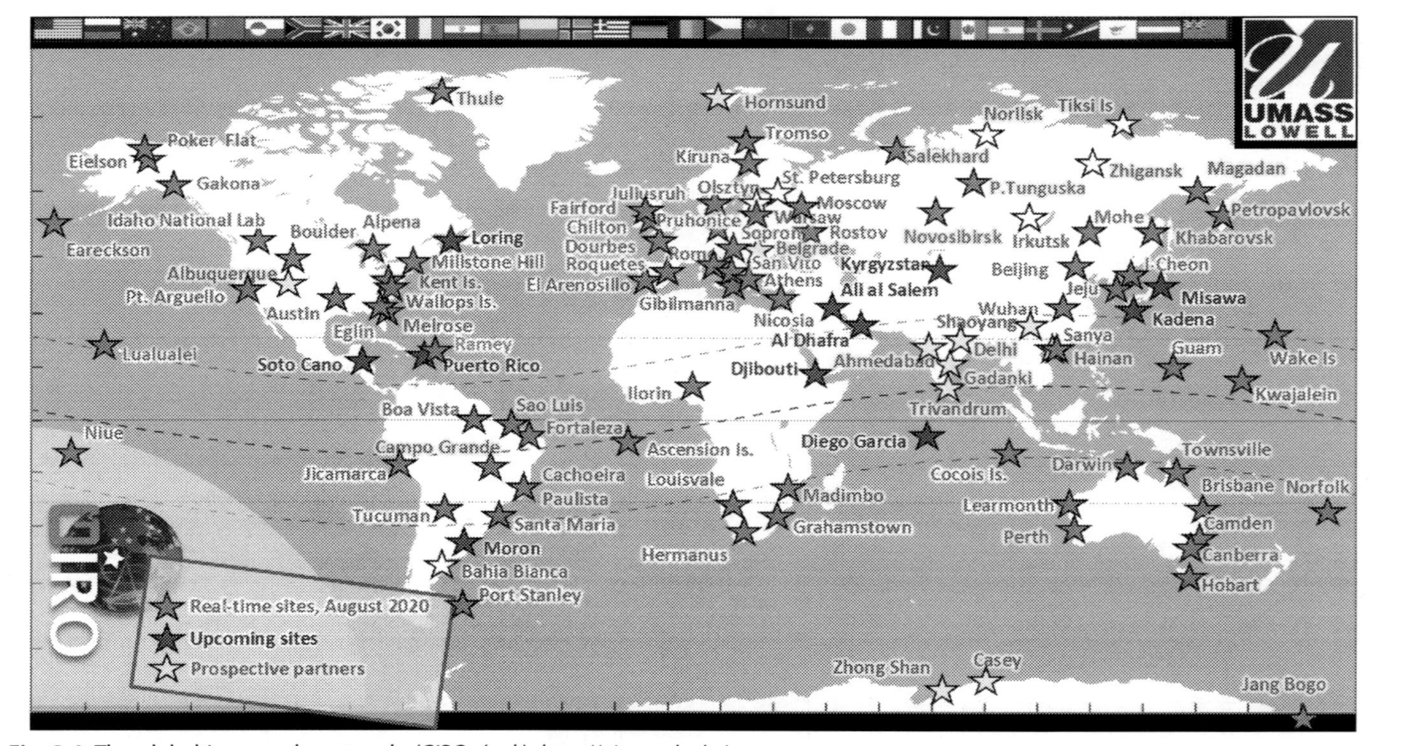

Fig. 3.4 The global ionosonde network. *(GIRO. (n.d.). http://giro.uml.edu.)*

many more. The characteristics of its LEO orbit are: close to polar with an inclination to the equator of 87.3 degree, altitude of 400 km, and a period of about 95 min. CHAMP provides extremely precise and accurate measurements of the Earth's magnetic and gravitational field. The GNSS-RO data that are suitable for processing are from the period May 2001 to October 2008. The mission ended on September 19, 2010 after 10 years, 2 months, 4 days, and 58,277 orbits.

The next satellite mission to provide GNSS-RO observations is the Constellation Observing System for Meteorology, Ionosphere, and Climate/Formosa Satellite Mission 3 (COSMIC/Formosat-3)—a system of six minisatellites with GNSS-RO receivers onboard for measuring the amplitude and phase of the GNSS signal. The satellite mission is a joint mission between the United States and Taiwan. The purpose of the COSMIC/ Formosat-3 mission is to observe the Earth and collect data to support synoptic, climatic, ionospheric, and gravitational analysis. The successful launch of the minisatellites into orbit took place in April 2006. The characteristics of the orbit are altitude 800 km and inclination 72 degree with respect to the Earth's equator.

TEC maps

As shown, the magnitude of ionospheric delay is proportional to the total electron content (TEC) along the signal path from satellite to receiver. TEC changes in space due to the spatial inhomogeneity of the ionosphere. Temporary changes are caused not only by the dynamics of the ionosphere, but also by sudden changes in the trajectory of transmission due to the movement of the satellite. When the signal comes from a small elevation angle, the trajectory of transmission through the ionosphere is much longer, so the corresponding delays can increase up to several meters at night and up to 50 m during the day. As the GNSS signal delay is greater at small elevation angles, this effect can be reduced by discarding measurements from low elevation angles. Since the impact of the ionosphere on the GNSS signals is frequency dependent, it can be filtered out using more than one frequency. For example, if receivers measure both GPS carrier frequencies L1 and L2, in order to eliminate the ionospheric delay the difference between the two frequencies can be used (see Eq. 3.5). A relatively simple analysis shows that the ionospheric delay changes inversely with the square of the wave frequency. Thus, the TEC can be computed using multiple satellites and reference networks of multiple frequency receivers. The state of the art in real-time global TEC maps is based on merging the GNSS measurements with empirical ionosphere models. In order to generate global TEC maps the ground-based GNSS TEC measurements are assimilated into an empirical TEC model. The Ionosphere Monitoring and Prediction Center (IMPC, n.d.), operated by the German Aerospace Center (DLR) at the Neustrelitz acquires real-time GPS data from different providers via the German Federal Agency for Cartography and Geodesy in Frankfurt. The GPS data are preprocessed in order to derive calibrated slant TEC (STEC) (Ciraolo, Azpilicueta, Brunini, Meza, & Radicella, 2007). As a next step, an empirical ionosphere model (Neustrelitz Total Electron Content Model, NTCM) is used as a background with GPS measurements updating the model coefficients. Subsequently, the calibrated STEC measurements are assimilated into NTCM (Jakowski, Mayer, Hoque, & Wilken, 2011) for generating the global TEC maps. Fig. 3.5 presents a TEC map from IMPC for a shell height of 400 km on October 4, 2020 at 8:15 UTC. The map is with temporal resolution of 15 min and horizontal resolution of 2.5 degree latitude × 5 degree longitude.

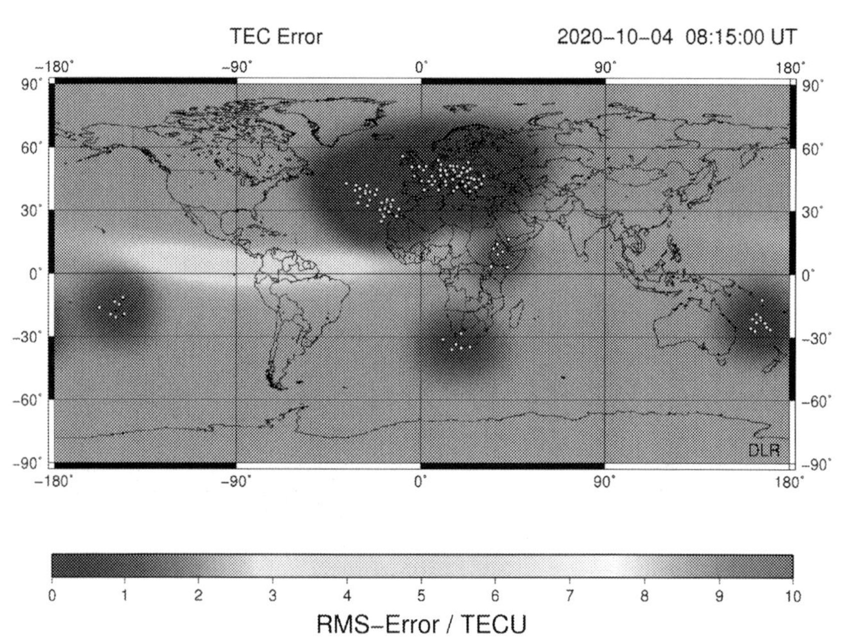

Fig. 3.5 Real-time global TEC map *(left)* and TEC error *(right)* at 8:15 UTC on October 4, 2020 from the Ionosphere Monitoring and Prediction Center. *(Real-time TEC (n.d.). https://impc.dlr.de/products/total-electron-content/near-real-time-tec/nrt-tec-global/.)*

Vertical profiles of electron density

The study of the vertical distribution of free electrons in the ionosphere is possible because satellite signals pass through different parts of the ionosphere and in addition the registration of the GNSS signals is on: (1) ground-based receivers and (2) LEO satellites such as SAC-C, CHAMP, or COSMIC/Formosat 3. GNSS radio occultation is a limb sounding technique (Fig. 3.6) for vertical refractivity profile retrieval of planetary atmospheres. The ray path bending and/or the phase of the radio wave are measured while approaching the Earth's surface in the limb sounding geometry (COSMIC, n.d.).

Fig. 3.7 presents the electron density profile (Ne) comparison between (1) CHAMP (IRO) instrument, (2) ionosonde (VS), and (3) a global empirical IRI model (Jakowski & Tsybulya, 2005). The comparison of IRO electron density observations with corresponding IRI model profile (Fig. 3.7 right) shows a systematic overestimation of the IRI-derived electron density above 250 km height and underestimation in the lower ionosphere (100–150 km). It is to be noted that the study reflects the IRI model skills as of 2005 and was a valuable input for model improvement. The IRO vertical profiles of electron density for a year with high solar activity (2002) and low solar activity (2006) are presented in Fig. 3.8 along with the TEC dependence of elevation angle. Visual comparison demonstrates the importance of the solar radiation for ionization at all altitudes and elevation angles.

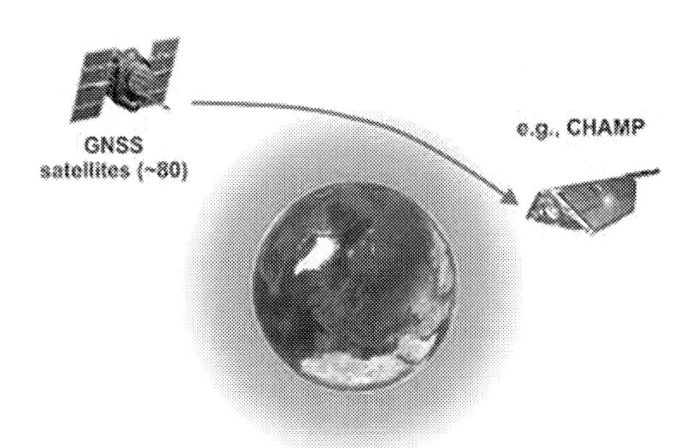

Fig. 3.6 Schematic presentation of GNSS limb sounding (GNSS-RO). *(GNSS radio occultation. (n.d.). http://www.tschmidt.eu/ro/ENG/main.html.)*

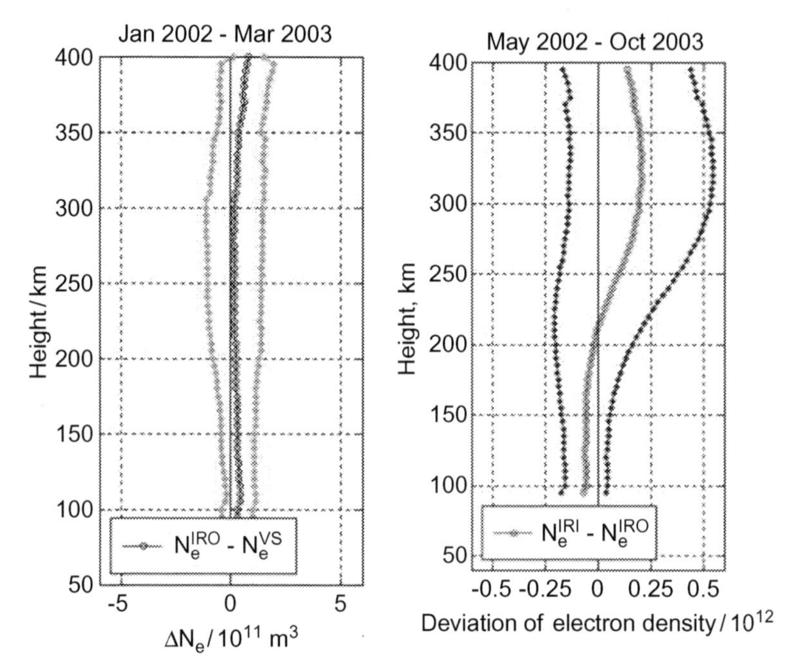

Fig. 3.7 Vertical electron density profiles derived from CHAMP and ionosonde station Juliusruh. The altitude-dependent mean difference *(blue and red lines; dark gray in print version)* and the corresponding standard deviation *(green lines; dark gray in print version)* are shown. *Left*: CHAMP—ionosonde and *right*: CHAMP vs IRI model. *(From Jakowski, N., & Tsybulya, K. (2005). Comparison of ionospheric radio occultation CHAMP data with IRI 2001. Advances in Radio Science, 2(G. 2), 275–279.)*

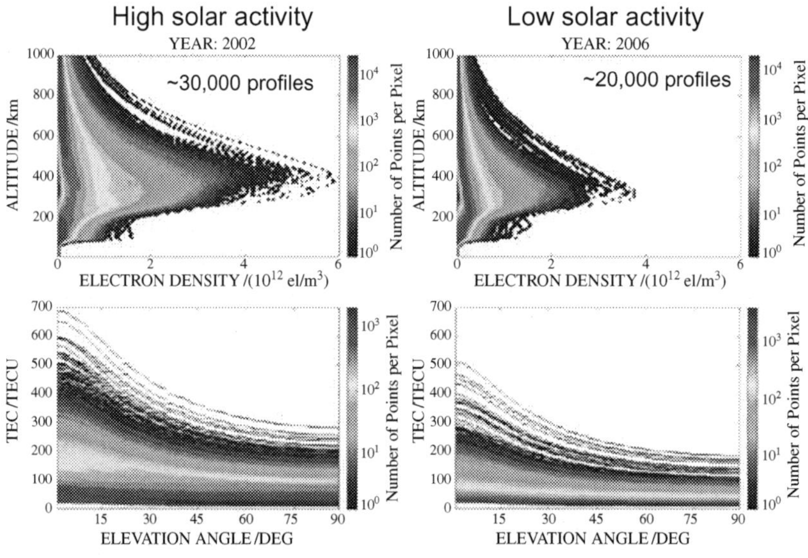

Fig. 3.8 CAMP-IRO vertical electron density profiles in 2002 *(top left)* and 2006 *(top right)* and TEC dependence on elevation angle in 2002 *(bottom left)* and 2006 *(bottom right)*. *(From Jakowski, N. (2005). Ionospheric GPS radio occultation measurements on board CHAMP. GPS Solutions, 9(2), 88–95. https://doi.org/10.1007/s10291-005-0137-7.)*

References

Bilitza, D. (2018). IRI the international standard for the ionosphere. *Advances in Radio Science, 16*, 1–11.

Ciraolo, L., Azpilicueta, F., Brunini, C., Meza, A., & Radicella, S. M. (2007). Calibration errors on experimental slant total electron content (TEC) determined with GPS. *Journal of Geodesy, 81*(2), 111–120. https://doi.org/10.1007/s00190-006-0093-1.

COSMIC. (n.d.). https://www.youtube.com/watch?v=qabMHoMyl1A&feature=youtu.be.

GIRO. (n.d.). http://giro.uml.edu.

GNSS Ionosphere explained. (n.d.). Https://Www.Youtube.Com/Watch?V=w-5Hl2b_wKE.

Hernández-Pajares, M., Aragón-Ángel, À., Defraigne, P., Bergeot, N., Prieto-Cerdeira, R., & García-Rigo, A. (2014). Distribution and mitigation of higher-order ionospheric effects on precise GNSS processing. *Journal of Geophysical Research: Solid Earth, 119*(4), 3823–3837. https://doi.org/10.1002/2013JB010568.

IMPC. (n.d.). https://impc.dlr.de/about/.

Ionogram. (n.d.). https://en.wikipedia.org/wiki/Ionosonde#Ionogram.

IRI. (n.d.). https://ccmc.gsfc.nasa.gov.

Jakowski, N., Mayer, C., Hoque, M. M., & Wilken, V. (2011). Total electron content models and their use in ionosphere monitoring. *Radio Science, 46*(6). https://doi.org/10.1029/2010RS004620.

Jakowski, N., & Tsybulya, K. (2005). Comparison of ionospheric radio occultation CHAMP data with IRI 2001. *Advances in Radio Science, 2*(G. 2), 275–279.

NASA Space weather. (n.d.). https://www.nasa.gov/mission_pages/rbsp/science/rbsp-spaceweather.html.

NOAA Aurora. (n.d.). https://www.swpc.noaa.gov/phenomena/aurora.

NOAA solar wind. (n.d.). https://www.nasa.gov/mission_pages/rbsp/science/rbsp-spaceweather.html.

NOAA Space weather. (n.d.). https://www.swpc.noaa.gov/impacts/space-weather-and-gps-systems.

SAC-C. (n.d.). https://www.nasa.gov/centers/goddard/pdf/110896main_FS-2000-11-012-GSFC-SAS-C.pdf.

Solar flares tracking. (n.d.). http://solar-center.stanford.edu/SID/activities/ionosphere.html.

CHAPTER 4

GNSS monitoring of the troposphere (GNSS-M)

When the GNSS signal passes through the atmosphere from the satellite to the ground receiver, it is subjected to various influences. The magnitude of these effects depends on the angle of the satellite above the horizon and the weather conditions. The atmosphere causes small but not negligible effects that include: (1) ionospheric group delay and ionospheric discharges (see Chapter 3), (2) group delay in the troposphere and stratosphere, and (3) atmospheric signal attenuation (Fig. 4.1).

Water vapor and the water cycle in the atmosphere

The troposphere can be considered to be composed of dry air and water vapor. The main gases composing the dry air are nitrogen, oxygen, argon, and carbon dioxide. Fig. 4.2A shows the vertical distribution of water vapor in the atmosphere. Half of the water vapor amount is concentrated in the lower 1.5 km of the atmosphere. The lower 5 km of the atmosphere contain 92% of the water vapor. The total condensed volume of water vapor in the atmosphere is 5.5 billion liters and it will cover the Earth evenly with a layer 25-mm thick, provided it is evenly distributed.

Water is the only substance on Earth that exists in nature in significant quantities in three phases: solid phase—ice, liquid phase—water, and gas—water vapor. Water vapor is one of the main gases in the troposphere (the lower 12 km of the Earth's atmosphere)—its amount varies from 0% to 7% of the volume of dry air, averaging about 4%. It is the most mobile form of water in the hydrological cycle of the Earth (Fig. 4.2B). Water vapor enters the atmosphere through evaporation from water bodies (oceans, seas, lakes, rivers), ice/snow cover and soil, as well as through evapotranspiration from vegetation. The condensation of water vapor in the atmosphere leads to the formation of clouds from which precipitation falls, i.e., water returns to the Earth's surface. Water vapor in the atmosphere has a relatively short life between 7 and 10 days, meaning that water in the atmosphere is

Global Navigation Satellite System Monitoring of the Atmosphere
https://doi.org/10.1016/B978-0-12-819152-1.00001-5

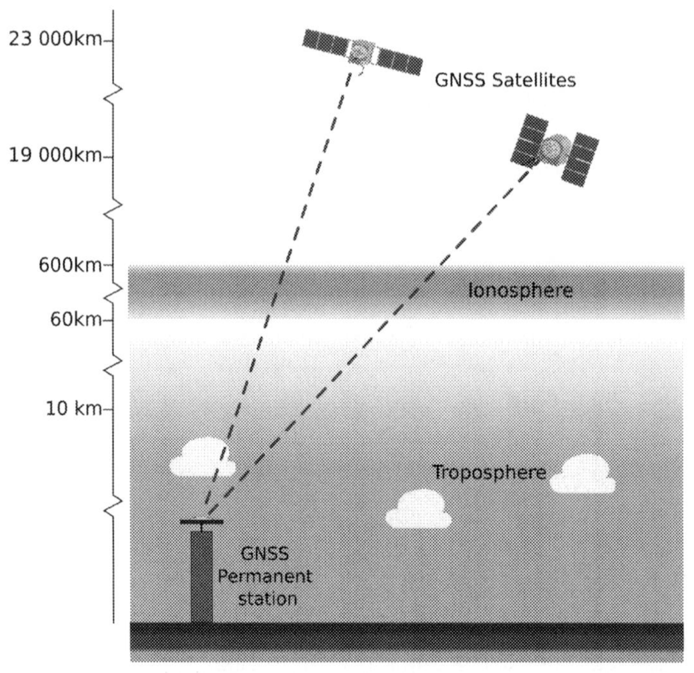

Fig. 4.1 Schematic presentation of GNSS signal propagation from satellite to the ground-based station. *(From Simeonov, T. (2021). Derivation and analysis of atmospheric water vapor and soil moisture from ground-based GNSS stations. In Technische Universität Berlin. Technische Universität Berlin.)*

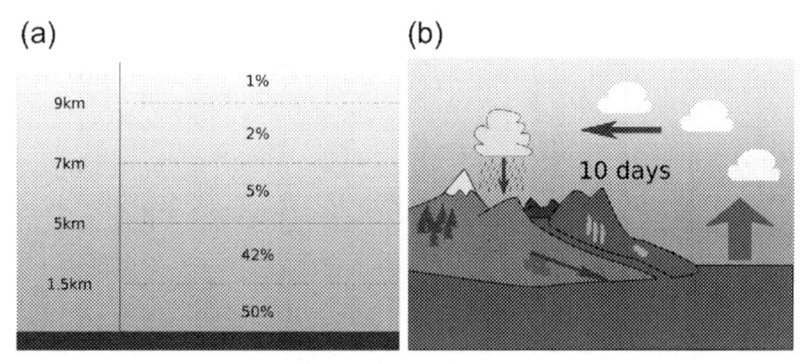

Fig. 4.2 (A) Vertical distribution of water vapor in the atmosphere. (B) Hydrological cycle. *(Courtesy Tzvetan Simeonov.)*

completely renewed about 45 times a year. Due to its high mobility, which includes vertical and horizontal transmission, and continuous phase transitions (evaporation/condensation), water vapor transfers large amounts of heat (hidden/latent heat) to the global energy redistribution. In addition, it is the main greenhouse gas in the atmosphere. That is why it is of particular importance for both the climate and the weather forecast. At the same time, due to the inhomogeneities in its distribution and to atmospheric dynamics and phase transitions, it is very difficult to measure.

Methods for measuring water vapor

Four independent methods for measuring water vapor in the atmosphere are discussed in this chapter (Table 4.1) (Bouma, 2002). Remote sensing of water vapor from space is one of the first applications of meteorological satellites. The first meteorological satellite mission, TIROS-1, launched in 1960 lasted 78 days and took pictures in the visible spectrum of the cloud cover of the Earth. The European Space Agency (ESA) launched the first satellite program in 1972. Seven satellites from the first-generation Meteosat-1 were deployed. The second-generation Meteosat (Meteosat-2) consists of 11 satellites in geostationary orbit. The image sensors onboard the Meteosat satellites are integrated in MVIRI—Meteosat Visible and Infrared Imager, operating in the thermal infrared (TIR), water vapor absorption (WV), and the visible range (VIS) (Desbois, Seze, & Szejwach, 1982). The Meteosat-2 temporal and spatial resolution provides 15-min time-resolution data from 12 different channels, including the 1-km high-resolution (HRV) visible channel with multiple spatial resolutions and multiple 2.5- and 5-km infrared channels (Schmetz et al., 2002). Using these satellite observations, water vapor is measured in channels with a frequency of 6.2 and 7.3 μm, focused at different levels in the troposphere, thus differentiating between the lower and the higher troposphere (Zinner, Mannstein, & Tafferner, 2008). The satellites effectively measure the absorption of water vapor from near-infrared radiation coming from the Earth's surface.

Table 4.1 Comparison of water vapor measurement methods.

	Radiosonde	Microwave radiometer	GNSS	Meteosat-2
Temporal resolution	Low	High	High	High
Spatial resolution	Low	Low	High	High
Vertical resolution	High	Low	Low	Low
All weather operation	Yes	No	Yes	No
Price	High	High	Low	High
In operation since	1950	1980	1990	1960

Modified from Bouma, H. R. (2002). Ground-based GPS in climate research. Chalmers University of Technology.

Radiosounding is performed with lighter than air balloons, filled with hydrogen or helium, to which a radiosonde is attached. Historically, radiosondes are the first system of meteorological sensors to measure remotely over a given point. The first attempts to measure upper atmospheric pressure and temperature were made in the 19th century. In 1892, a balloon with temperature and pressure measuring devices (thermograph and barograph) was used for the first time. With the start of regular measurements of meteorological elements, for the first time it was shown that the temperature decreases to a certain altitude and then remains constant. This is how the tropopause was discovered. In the beginning of the 1930s, the radiosonde was used for the first time, and after 1950 it became the most common method of studying the atmosphere. With the discovery and proof of important properties of the atmosphere, the method was approved by the World Meteorological Organization (WMO). It is widely used to measure the vertical distribution of temperature, pressure, and humidity in the atmosphere and to determine wind speed and direction. Regular sounding is usually performed once or twice a day at 00 and/or 12 UTC from a global network of aerological stations. The distance between the stations where the sounding is performed is 250 km or more. Fig. 4.3 shows the global map with the average number of observations in 24 h for the month of May 2021 from operational radiosonde stations.

The other two methods for measuring the amount of water vapor are based on the fact that the propagation of electromagnetic waves is influenced by the composition of the atmosphere. The microwave radiometry is based on the influence of water vapor, water, and oxygen on the propagation of waves. Microwave radiometers measure atmospheric emissions at 21 and 31.4 GHz. The brightness temperature corresponding to the intensity of this radiation is directly related to the amount of water vapor. Microwave radiometers provide a high temporal resolution, but the measurements are obstructed by liquid water droplets and are very costly, so they are mainly used for research.

GNSS meteorology

Brief history

The application of GNSS in meteorology was proposed in 1992 by Bevis et al. (1992). Since 1996, six GNSS meteorology projects have been funded in Europe (Fig. 4.4). In the first project (MAGIC), Haase, Ge, Vedel, and Calais (2003) and Vedel, Huang, Haase, Ge, and Calais (2004) studied the

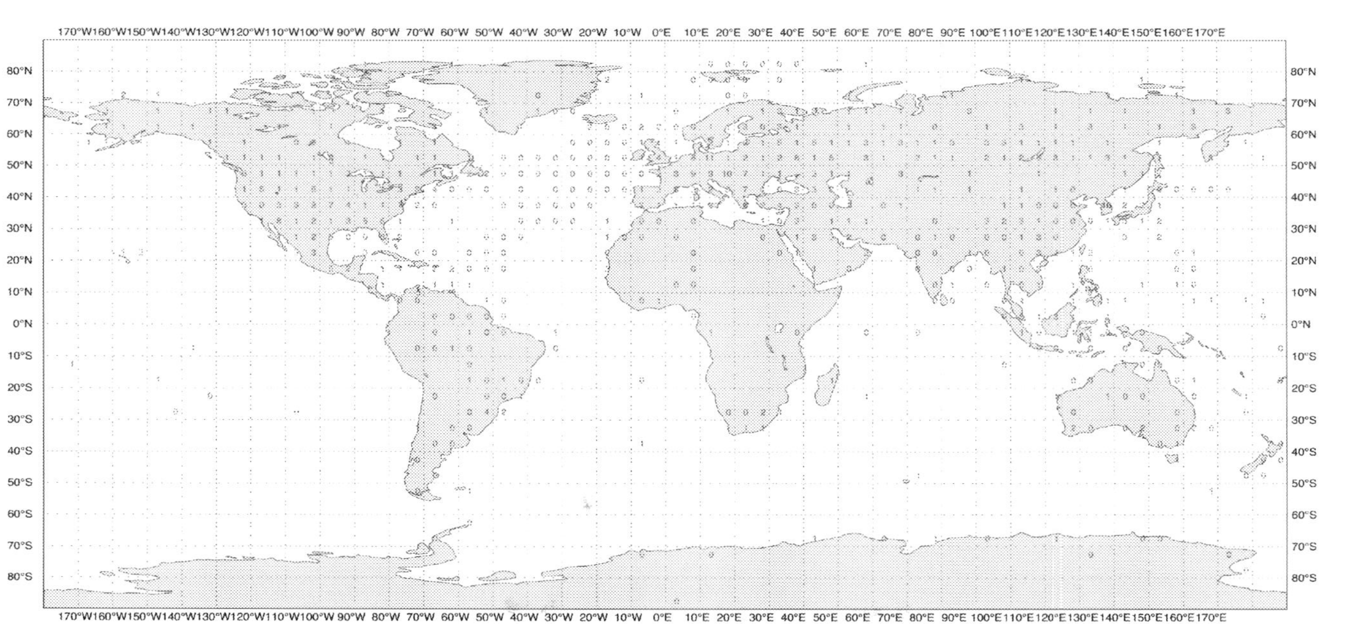

Fig. 4.3 Average number of radiosonde observations in 24 h for May 2021. *((N.d.). Retrieved from https://www.weather.gov/jetstream/radiosondes (Accessed 6 July 2021).)*

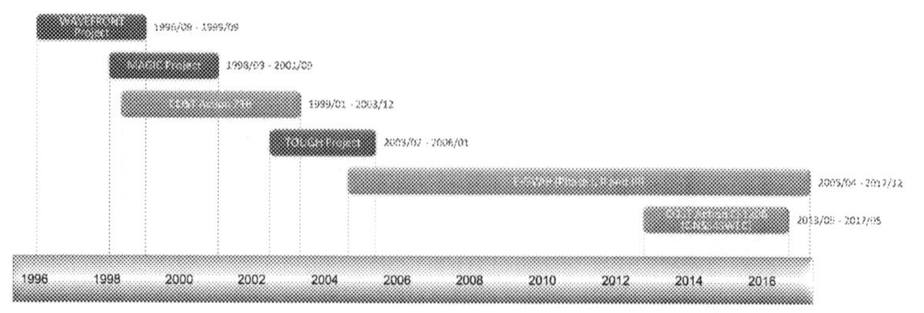

Fig. 4.4 European GNSS Meteorology projects. *(From Guerova, G., Jones, J., Douša, J., Dick, G., Haan, S. de, Pottiaux, E., Bock, O., Pacione, R., Elgered, G., & Vedel, H. (2016). Review of the state of the art and future prospects of the ground-based GNSS meteorology in Europe. Atmospheric Measurement Techniques, 9(11), 5385–5406.)*

application of GNSS meteorology in the Western Mediterranean. The second project was the scientific network COST 716 "Use of terrestrial GPS in climate and numerical weather forecasting" (COST-716) (Elgered, Plag, van der Marel, Barlag, & Nash, 2005). As part of COST-716, real-time GNSS observations from more than 200 European stations were provided for experiments with numerical weather forecasting models used by five national meteorological services (German Meteorological Service—DWD, Swiss Meteorological Service—MeteoSwiss, British Meteorological Service—Metoffice, Swedish Meteorological Service—SMHI, Danish Meteorological Service—DMI). In 2003, the TOUGH project was launched. TOUGH is a continuation of COST-716 and further develops the methods for the assimilation of GNSS observations in numerical models for weather forecasting (see "Real-time monitoring with GNSS IWV" section). The EGVAP operational service started in 2005 as a continuation of COST-716 and TOUGH and from 2019 entered its fourth phase. Its purpose is to provide in real time the GNSS data required for operational weather forecasting. Currently, more than 2500 GNSS stations (Fig. 4.5) are processed by the 19 GNSS Analysis Centers and provided to EGVAP members. In 2013, a second scientific network COST-1206 "Advanced Global Navigation Satellite Systems tropospheric products for monitoring severe weather events and climate (GNSS4SWEC)" was launched (Jones et al., 2019).

GNSS meteorology—Integrated water vapor

Fig. 4.6 shows the path of three GNSS signals from the satellite to the ground-based receiver. These slanted paths [slant tropospheric delay (STD)] are combined and the common path in the zenith direction called

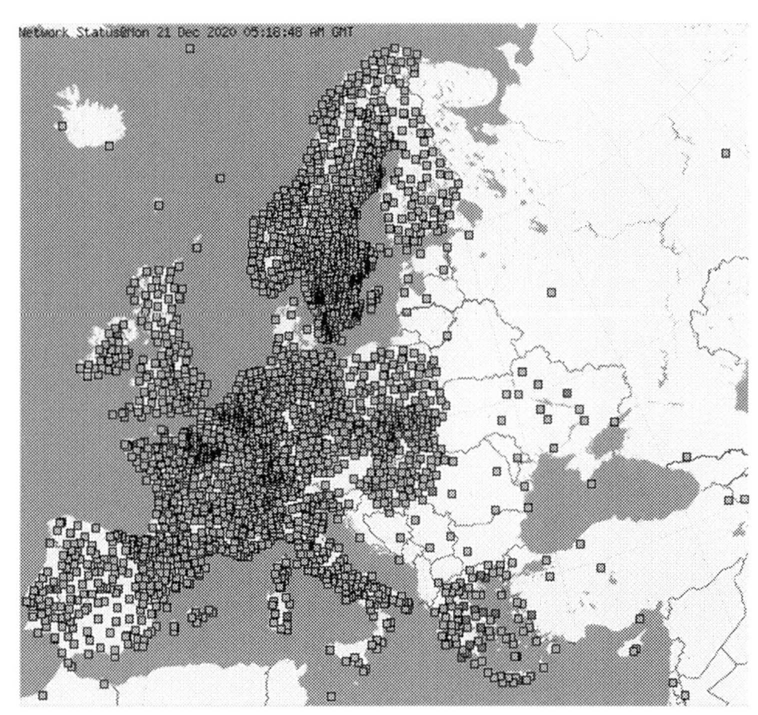

Fig. 4.5 *Green color* (dark gray in print version) represents the stations with information from 16 to 17 UTC on December 21, 2020, yellow (light gray in print version) those from 13 to 16 UTC, and *red color* (dark gray in print version) represents those from 13 to 19 UTC. *((N.d.). Retrieved rom http://projects.knmi.nl/egvap/validation/europe.html (Accessed 21 December 2020) EGVAP network of GNSS ground stations. Source: http://projects.knmi.nl/egvap/validation/europe.html).*

the zenith total delay (ZTD) is calculated. At GNSS operating frequencies, oxygen is the main cause of signal attenuation as it passes through the atmosphere, but this effect is negligible. The group delay resulting from tropospheric refraction is a major source of GNSS error (described in detail in Chapter 2). There are two sources of group delay. The first and greater delay is due to dry air, mainly nitrogen and oxygen, and is called the zenith hydrostatic delay (ZHD). ZHD is between 2.0 and 2.2 m at sea level and varies with temperature and atmospheric pressure (upper panel of Fig. 4.7). As it can be seen from Fig. 4.7, the change over time in the ZHD is less than 1% of the mean over several hours. The second delay is the result of water vapor present in the troposphere and is called the zenith wet delay (ZWD). ZWD has low absolute value (1–80 cm) but is characterized by high

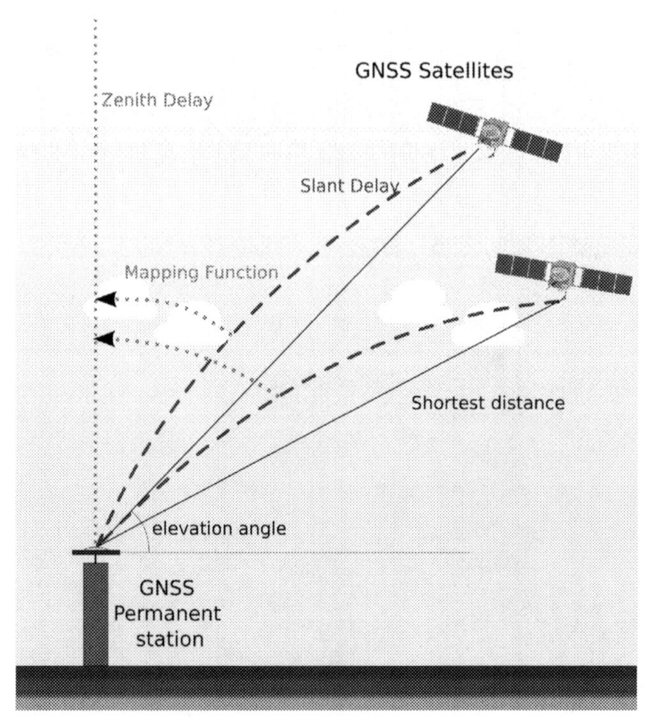

Fig. 4.6 GNSS slant delays *(in blue; dark gray in print version)* mapping to the zenith direction *(in red; dark gray in print version). (From Simeonov, T. (2021). Derivation and analysis of atmospheric water vapor and soil moisture from ground-based GNSS stations. In Technische Universität Berlin. Technische Universität Berlin.)*

temporal variability (10%–20% for several hours) and this makes it difficult to estimate. The vertically integrated water vapor (IWV) is estimated from the ZWD. GNSS meteorology concept is presented in an animated video (GNSS Meteorology Explained, n.d.).

Ground pressure (p_s) and temperature (T_s) are required to calculate the integrated water vapor from ZWD. They can be from both observations and a numerical model. IWV is calculated as suggested by Bevis et al. (1992) and Davis, Herring, Shapiro, Rogers, and Elgered (1985). First, the zenith water delay (ZWD) is calculated as the difference between the zenith total delay and the zenith hydrostatic delay:

$$ZWD = ZTD - 0.0022768 \frac{p_s}{1 - 0.00266 \cos(2\theta) - 0.00028h} \tag{4.1}$$

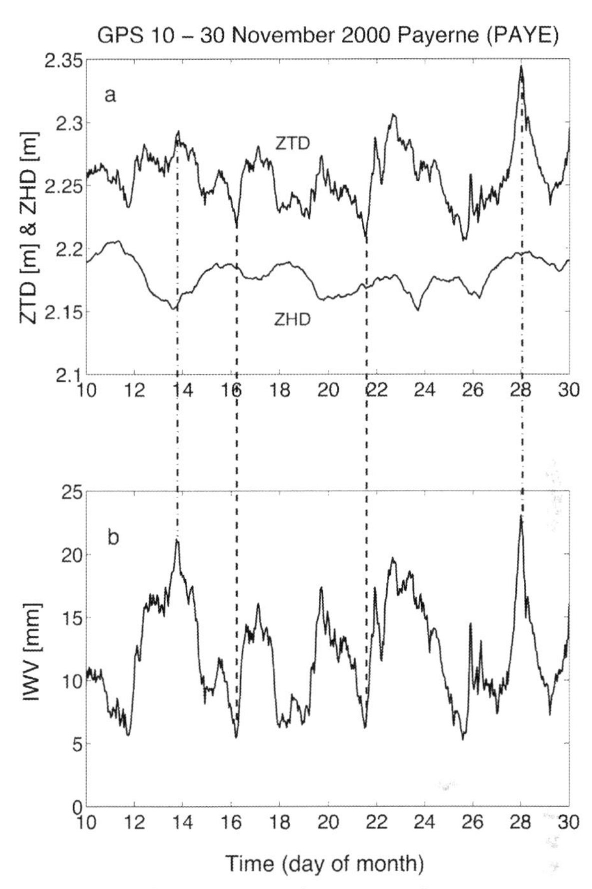

Fig. 4.7 ZTD, ZHD *(top)* and IWV *(bottom)* from November 10 to 30, 2000 at Payerne, Switzerland. *(From Guerova, G., Brockmann, E., Quiby, J., Schubiger, F., & Matzler, C. (2003). Validation of NWP mesoscale models with Swiss GPS network AGNES. Journal of Applied Meteorology, 42(1), 141–150.)*

and IWV:

$$IWV = \frac{10^6}{(k_3/T_m + k_2')\,R_v} ZWD \tag{4.2}$$

where $T_m = 70.2 + 0.72\ T_s$, k_2' and k_3 are constants, empirically determined for specific regions (Bevis et al., 1992). $R_v = 461.51 \frac{J}{kg \cdot K}$ is the gas constant for water vapor, T_m is the mean temperature in the troposphere, h is the height, and θ is the latitude variation of gravitational acceleration at a given location.

As surface pressure is a critical element if there is a difference between the height of the GNSS station and the meteorological station it has to be taken into account by using the polytropic barometric formula (Sissenwine, Dubin, & Wexler, 1962):

$$P_g = P_m \left(\frac{T}{T - L(H_g - H_m)} \right)^{\left(\frac{g_0 M_0}{R L} \right)} \tag{4.3}$$

where p_g is the pressure at the GNSS station, p_m is the pressure at the meteorological station, T is the surface temperature at the meteorological station, $L \approx 6.5 \frac{K}{km}$ is the temperature lapse rate, H_m is the altitude above sea level of the meteorological station, H_g is the altitude above sea level of the GNSS station, g_0 is the gravitational acceleration, $M_0 = 28.9644 \frac{g}{mol}$ is the molar mass of air, and $R = 8.31432 \frac{N \cdot m}{mol \cdot K}$ is the universal gas constant.

Integrated water vapor—Pressure and temperature

An important element in the calculation of IWV is the surface pressure (formula 4.1) and temperature (formula 4.2). In Fig. 4.8, the IWV for two GNSS stations in Switzerland, Payern (600 m a.s.l., blue bars) and Andermatt (2300 m a.s.l., purple bars) is shown. In addition, IWV is computed for pressure deviations of ± 2 hPa (upper panel of Fig. 4.8) and temperature ± 2 K (lower panel of Fig. 4.8). The absolute values are provided in the left column, while the deviations of the integrated water vapor are in the right column. For GNSS stations Payerne and Andermatt, pressure deviations of: (1) ± 1 hPa result in integrated water vapor changes of ± 0.35 kg/m^2 and (2) ± 2 hPa result in changes of ± 0.70 kg/m^2. For Payerne station, temperature deviations of: (1) ± 1 K result in integrated water vapor changes of ± 0.06 kg/m^2 and (2) ± 2 K result in changes of ± 0.12 kg/m^2. For Andermatt station, temperature deviations of ± 2 K lead to changes in the integrated water vapor of ± 0.06 kg/m^2. It can be concluded that accurate surface pressure determination is a leading factor in obtaining integrated water vapor with an accuracy higher than ± 1 kg/m^2. Thus, the pressure should be interpolated to the correct altitude of the GNSS station using the barometric formula 4.3.

IWV example GNSS and radiosonde

A comparison of integrated GNSS water vapor and radiosonde (RS) is presented in Fig. 4.9. The measurements are performed at Payerne station,

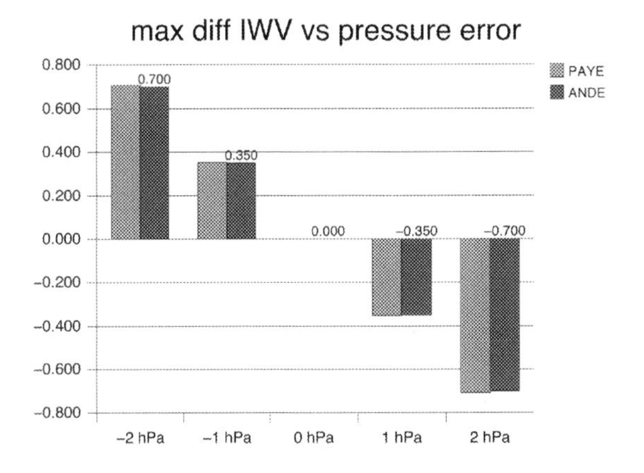

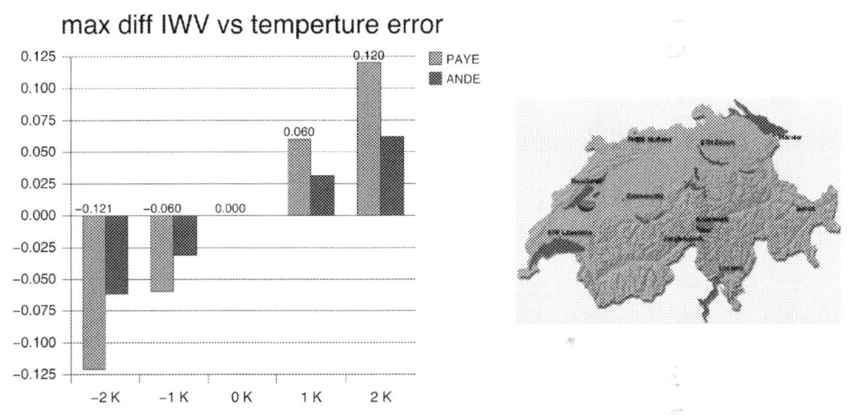

Fig. 4.8 Contribution of pressure *(upper panel)* and temperature *(bottom panel)* to the IWV error at Payerne *(blue bars)* and Andermatt *(purple bars)* Switzerland. *((N.d.). Courtesy Guergana Guerova.)*

Switzerland, where the radiosounding station has been collocated with GNSS station since 2000. The period January 5–30, 2001 shows a very good agreement between GNSS IWV (red line; dark gray in print version) and RS observations (green squares; dark gray in print version), with the exception of the period from 00 UTC on January 14 to 12 UTC on January 16, 2001. During this 3-day period, it was found that the IWV measurement was a factor of two higher than the radiosonde measurement (Guerova, Brockmann, Quiby, Schubiger, & Matzler, 2003). The meteorological conditions during the period were characterized by low-level clouds and temperature inversion. It was found that during this period the humidity sensor of the radiosonde has a delayed response time after passing through the low-level clouds and this is the reason for the obtained high values.

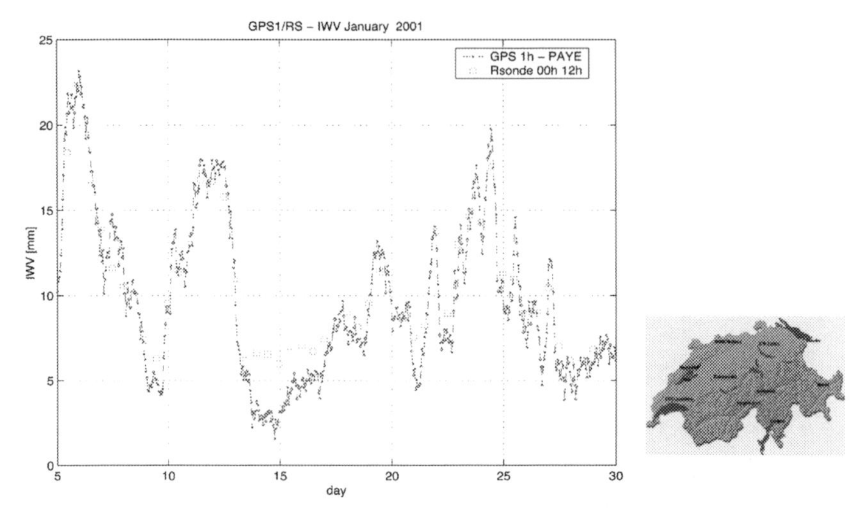

Fig. 4.9 GNSS integrated water vapor *(red line; dark gray line in print version)* and aerological drilling *(green squares; dark gray squares in print version)* for the period January 5–30, 2001. Aerological and GNSS stations are in Payerne, Switzerland *(right panel)*. *(From Guerova, G. (2003).)*

Guerova, Brockmann, Schubiger, Morland, and Mätzler (2005) reported a small positive bias of $0.3\,\mathrm{kg/m^2}$ between GNSS and radiosonde IWV at Payerne for the period from January 2001 to June 2003. The day-night observation relationship was studied, similar to that by Ohtani and Naito (2000), and the mean deviation (RS minus GNSS) was calculated at 00 UTC (upper panel of Fig. 4.10) and at 12 UTC (lower panel of Fig. 4.10). An average difference found at 00 UTC was negative $-0.4\,\mathrm{kg/m^2}$, and that at 12 UTC was $+0.9\,\mathrm{kg/m^2}$. The main contribution to the positive deviation of 1.5 and $0.9\,\mathrm{kg/m^2}$ at 12 UTC was from 2002 and 2003. The relative difference at 12 UTC in 2003 was 3.5%. It can be summarized that for the observation at 12 UTC the radiosonde tends to increase the values of the integrated water vapor. For the period 2002–2003, the comparison of the observations from GNSS and RS at 00 UTC leads to a difference of −2.3%, which shows a tendency for decrease in the integrated water vapor in the midnight RS observation.

Real-time monitoring with GNSS IWV

An example for IWV evaluation of the local model analysis (LMa), operated by the German Weather Service (DWD), with near real-time GNSS data for

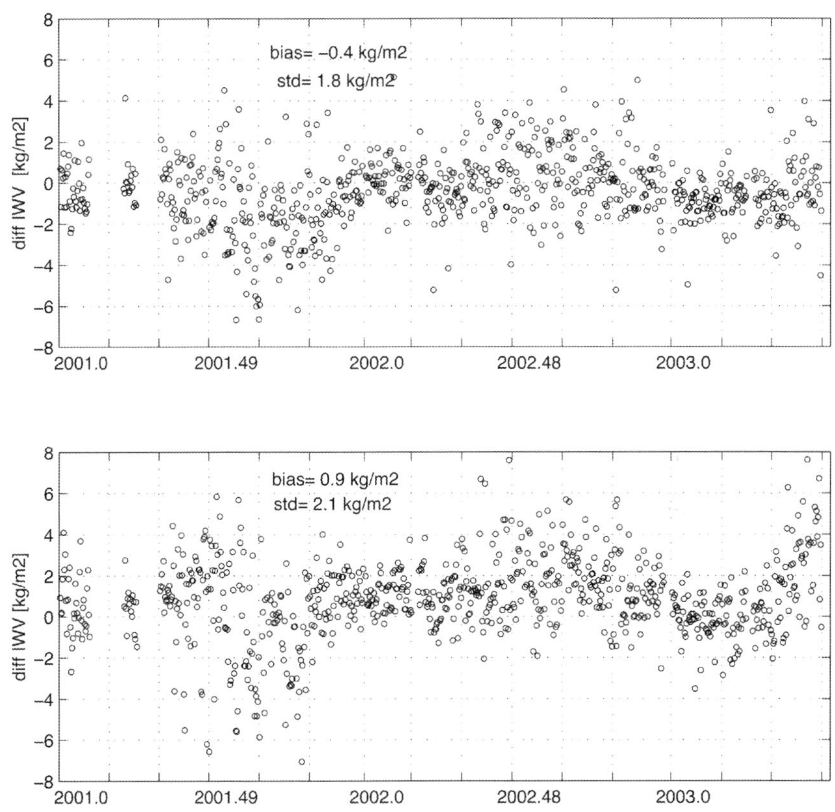

Fig. 4.10 IWV difference (RS minus GNSS) at 00 UTC *(upper panel)* and 12 UTC *(lower panel)* for Payerne station, Switzerland during the period 2001–2003. *(From Guerova, G., Brockmann, E., Schubiger, F., Morland, J., & Mätzler, C. (2005). An integrated assessment of measured and modeled integrated water vapor in Switzerland for the period 2001–03. Journal of Applied Meteorology, 44(7), 1033–1044.)*

April 24, 2001 is provided in this section. The LMa IWV field is plotted in Fig. 4.11 as a color map. The GNSS IWV is color coded within the black circles. A visual IWV comparison of the LMa and GNSS shows very good agreement in the central and eastern part of the map. However, in the western part of the map (large red circle; dark gray in print version; Fig. 4.11), the LMa IWV has a maximum of more than 20 kg/m^2, while the GNSS IWV value is 10 kg/m^2. As reported by Guerova et al. (2003), the reason for this large difference is the assimilation in the LMa of a radiosonde observation. The radiosonde profile was assimilated in the model and assimilation of only one incorrect temperature observation, with 10°C higher than the model,

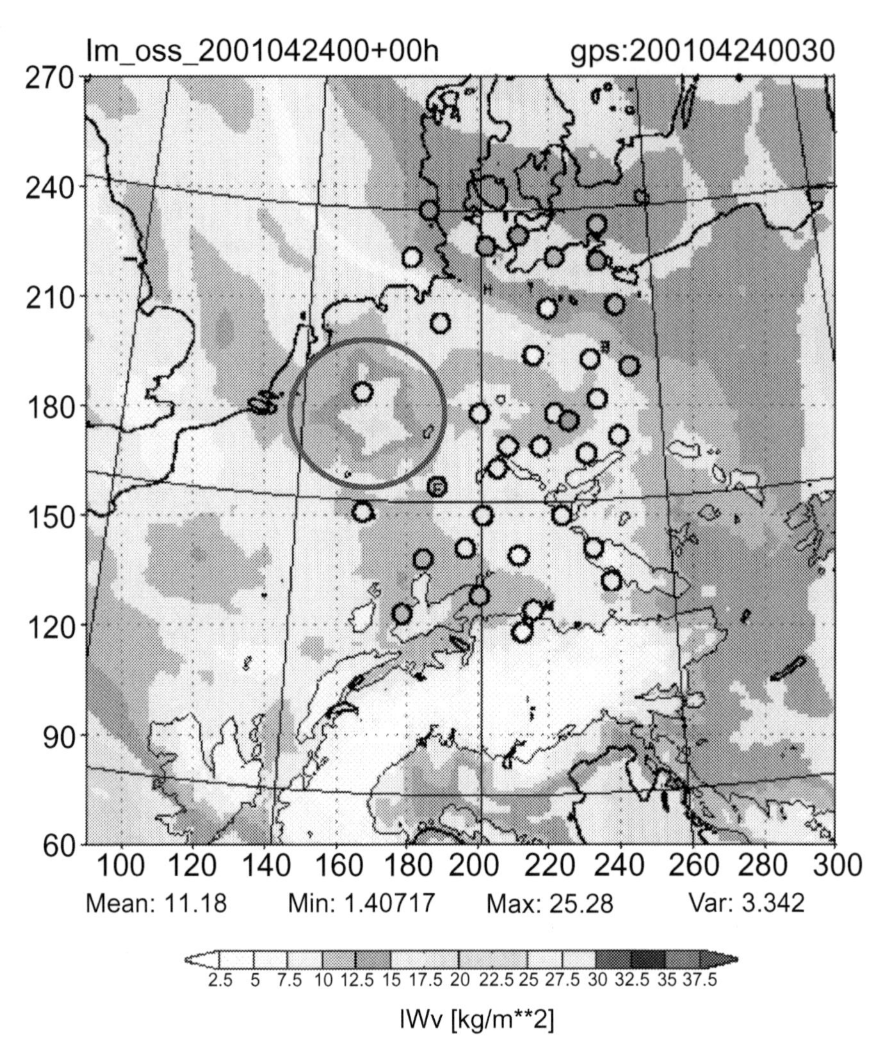

Fig. 4.11 IWV from LMa (color map) and the GNSS *(color in black circles)*. The *red circle* (dark gray in print version) marks the region with the large difference between GNSS and LMa. *(From Guerova, G. (2003).)*

resulted in doubling of model IWV. It is to be noted that this single incorrect observation was spread by the model in the surrounding area with decreasing weight resulting in the onion-shell shape with green (encirled dark gray in print version) at the center (20–25 kg/m^2) and orange (light gray area adjacent to the large circle in print version) at the periphery (10–15 kg/m^2). This example demonstrates the added value of GNSS near real-time data to the quality control procedures applied to radiosonde observations assimilated in the model. Chapter 5 is dedicated to GNSS and numerical weather modeling.

GNSS monitoring: precipitation, hail, and thunderstorm

Evaluation of humidity distribution is crucial in order to determine the possibility of convection, heavy precipitation, hail, and thunderstorms. Storms usually develop where humidity is already high or where some mechanism makes it increase. The high spatiotemporal availability of GNSS-derived water vapor is well suited for monitoring the humidity changes. In this section, four examples for GNSS monitoring of precipitation in Portugal, hail in Argentina, cloud to ground lightning strokes in thunderstorms in Japan, and thunderstorm days in Bulgaria and Poland are discussed.

Benevides, Catalao, and Miranda (2015) investigated the relationship between precipitation and IWV in the Lisbon region, Portugal. Precipitation probability is plotted as a function of the maximum IWV (Fig. 4.12A), the maximum IWV increase for 6 h (Fig. 4.12B), and the maximum IWV change rate (Fig. 4.12C). The precipitation probability increases with the maximum value of IWV and reaches above 70% for IWV higher than 42 kg/m^2 or IWV maximum increase above 15 kg/m^2. The precipitation probability and the IWV change rate are shown to be linearly dependent (Fig. 4.12C). The linearity in the dependence between precipitation rate and IWV change rate is demonstrated only on a regional scale.

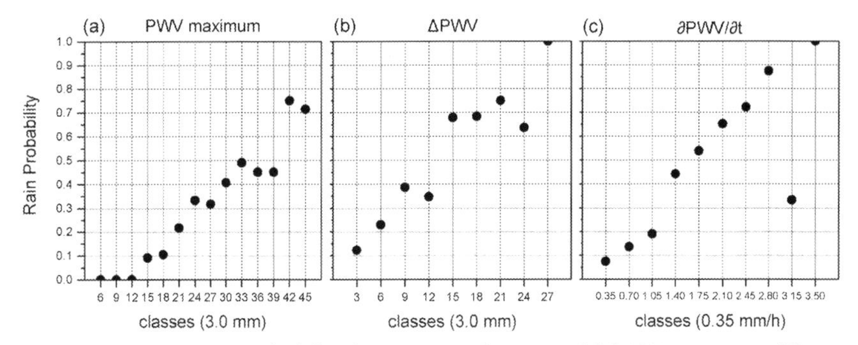

Fig. 4.12 Precipitation probability for 2012 as a function of (A) IWV maximum, (B) IWV maximum increase (△ PWV), and (C) IWV change rate (∂PWV/∂t) *(From Benevides, P., Catalao, J., & Miranda, P. (2015). On the inclusion of GPS precipitable water vapor in the nowcasting of rainfall. Natural Hazards and Earth System Sciences, 15(12), 2605–2616.)*

Calori et al. (2016) investigated the link between IWV and hail in a region of Mendosa province in Argentina over a 45-day period (January–February 2010). Fig. 4.13 presents the IWV anomaly and the registered hail (red bar; dark gray boxes above the horizontal axis line in print version) and precipitation without hail (blue bars; dark gray boxes below the horizontal axis line in print version). A visual inspection of the figure indicates that the positive IWV anomalies are associated with hail registrations, while the negative IWV anomalies result only in precipitation without hail. This result indicates that GNSS anomalies show a potential for the prediction of possible hail events.

Inoue and Inoue (2007) investigated the time lag between IWV maximum and cloud to ground lightning strokes in Kanto District, Japan. They found that 40% of the cloud to ground strokes are preceded by IWV maximum of 15–30 min. As can be seen in Fig. 4.14, the time lag between IWV and cloud to ground strike maximum is up to 4 h. Both IWV and the 30-min IWV increment are reported to have large values within 1 h before the stroke. These results demonstrate the applicability of GNSS-derived IWV in short-range forecasting of lightning activities. Forecasting lightning and thunderstorms are of major importance for safety procedures in aviation.

Guergana Guerova, Dimitrova, and Georgiev (2019) studied the thunderstorm activity in Sofia Plain, Bulgaria over the period from May to

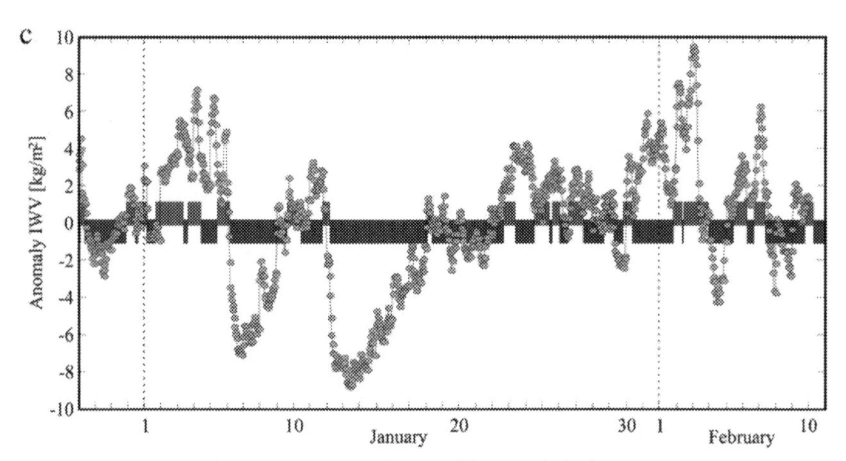

Fig. 4.13 IWV anomaly *(gray line with dots)* and hail *(red; dark gray in print version)* and no hail *(blue; dark gray in print version)* precipitation days. *(From Calori, A., Santos, J. R., Blanco, M., Pessano, H., Llamedo, P., Alexander, P., & de la Torre, A. (2016). Ground-based GNSS network and integrated water vapor mapping during the development of severe storms at the Cuyo region (Argentina). Atmospheric Research, 176, 267–275.)*

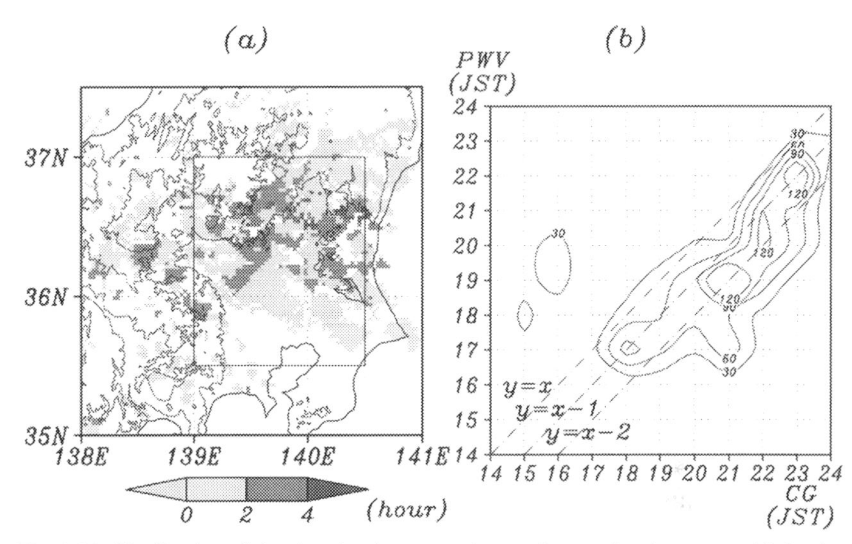

Fig. 4.14 Distribution of the time lag between the maximum cloud to ground lightning stroke and maximum IWV. *(From Inoue, H. Y., & Inoue, T. (2007). Characteristics of the water-vapor field over the Kanto district associated with summer thunderstorm activities. SOLA, 3, 101–104.)*

September 2010–2015. As can be seen from the box and whiskers plot in Fig. 4.15, the daily mean IWV separates well the days with thunderstorms (TH) and without thunderstorms (NTH). A stepwise discriminant analysis was applied to derive classification functions based on (1) IWV, (2) eight instability indices derived from radiosonde data, and (3) combination of IWV and (2). The skill scores for probability of detection (POD), false alarm ratio (FAR), critical success index (CSI), and true skill statistic (TSS) give the best results when IWV is combined with instability indices as shown in Fig. 4.16. Evaluation of the monthly classification functions was carried out using an independent sample period (2017–2018) and it was confirmed that the best scopes are obtained by the classification function combining IWV and instability indices for all months from May to September.

A machine learning approach was used by Łoś, Smolak, Guerova, and Rohm (2020) to study the thunderstorm predictive potential of the synergy between IWV and GNSS tomography-based vertical wet refractivity profiles (Nw) in Poland. The results show that wet refractivity below 7 km, namely, at altitudes 1.75, 2.25, 3.75, and 6.75 km, and 1–2 h ahead of the thunderstorm are the five most important features. IWV 1 and 3 h ahead and to the west of the thunderstorm are number 6 and 7 of the most significant parameters with predictive potential (Fig. 4.17).

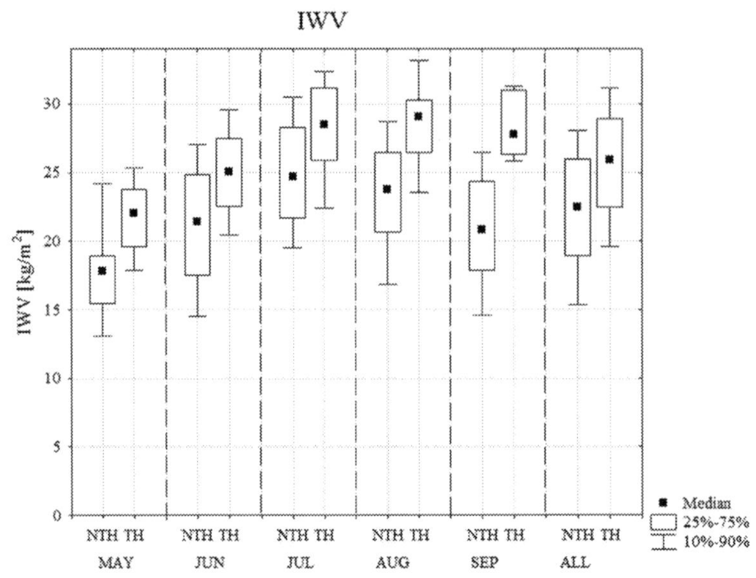

Fig. 4.15 Box and whisker plot of IWV median *(black square)* and 10, 25, 75, and 90 percentiles for TH and NTH days, from May to September 2010–2015. *(From Guerova, G., Dimitrova, T., & Georgiev, S. (2019). Thunderstorm classification functions based on instability indices and GNSS IWV for the Sofia Plain. Remote Sensing, 11(24). https://doi.org/10.3390/rs11242988.)*

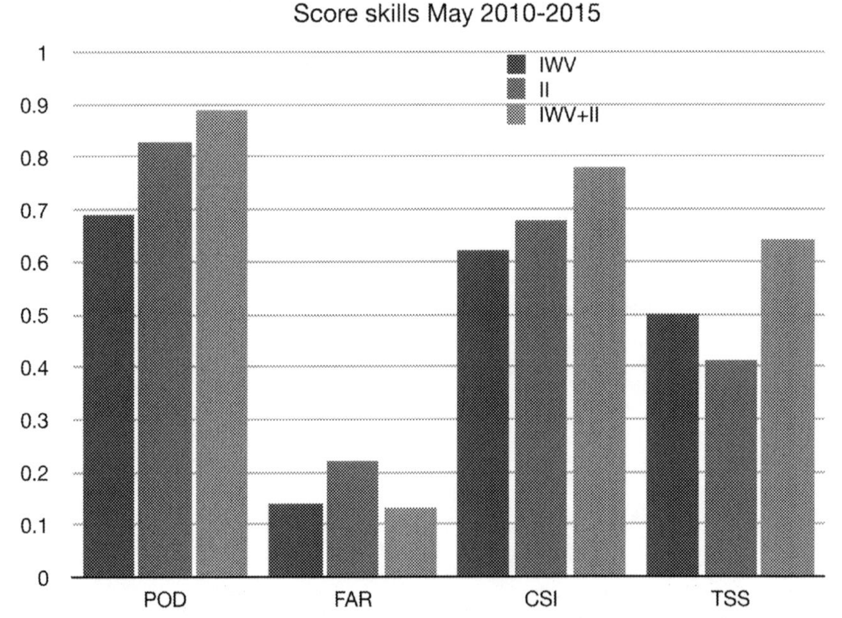

Fig. 4.16 Monthly skill scores for classification functions based on IWV *(left)*, II *(center)*, and IWV and II *(right)* for May 2010–2015. *((N.d.). Courtesy Guergana Guerova.)*

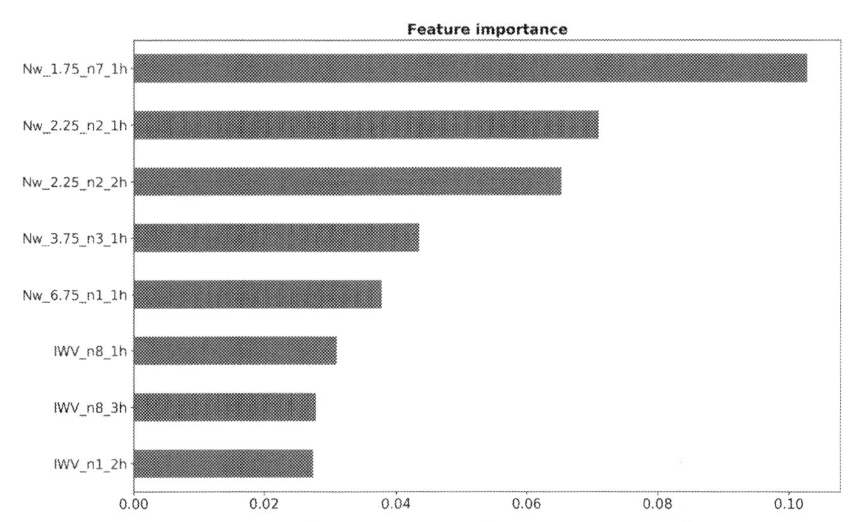

Fig. 4.17 Features important for thunderstorm location prediction for Poland. *(From Łoś, M., Smolak, K., Guerova, G., & Rohm, W. (2020). GNSS-Based Machine Learning Storm Nowcasting. Remote Sensing, 12(16). doi: 10.3390/rs12162536.)*

GNSS monitoring service in Bulgaria

In 2017, a new collaboration in the Balkan-Mediterranean region was initiated between Bulgaria, Cyprus, and Greece in the framework of the Transnational Cooperation Programme Interreg with a regional project "BalkanMed real time severe weather service" (BeRTISS) (Haralambous et al., 2018). The BeRTISS project objective is to establish a pilot transnational severe weather service by exploiting GNSS tropospheric products and enhance the safety, the quality of life, and environmental protection in the region. The Bulgarian Hail Suppression Agency and the Sofia University "St. Kliment Ohridski" are the Bulgarian partners in the BeRTISS project. In Bulgaria, severe weather events, such as intense precipitation, hail, and thunderstorms, are common in the summer months and are associated with large economic losses. For example, a hail storm of July 8, 2014 in Bulgarian capital Sofia caused widespread damage to over 50,000 vehicles and over 123 million euro in insured losses for an approximate duration of 2h. BeRTISS severe weather service in Bulgaria targets operational use in hail suppression activities in Bulgaria. The first step was the establishment of a network of 12 ground-based GNSS stations in Bulgaria in 2018 as a part of the BeRTISS project. The Bulgarian Hail Suppression Agency (HSA) operates nine GNSS stations (Fig. 4.18) and three stations are operated by the Sofia University

Fig. 4.18 Map of Bulgaria with location of BeRTISS GNSS stations *(blue pointer; dark gray in print version)* and collocated GNSS and radar station *(red pointer; dark gray in print version). (Bulgarian Integrated NowCAsting tool (BINCA). (2021). https://binca-bg.eu/.)*

(SU). The station's geographical coverage is in southwestern and central Bulgaria. The GNSS stations deliver hourly data streams to the operational server at the Sofia University (Physon). The GNSS observation data flow and processing is presented in Fig. 4.19. The Sofia University GNSS Analysis Center (SUGAC) processes 12 Bulgarian stations and 31 European stations. The processing is done using Bernese software v5.2 and a double-difference strategy. For GNSS data pre- and postprocessing, the TropNET

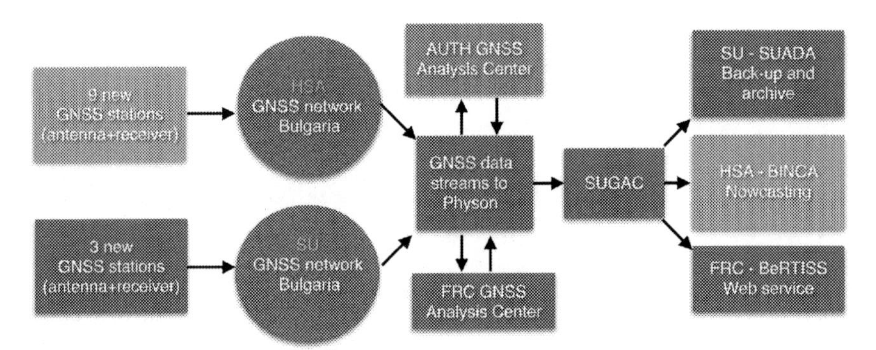

Fig. 4.19 GNSS observation data flow in Bulgaria and data exchange with BeRTISS partners in Cyprus (FRC) and Greece (AUTH). *(From Guerova, G., Dimitrova, T., & Georgiev, S. (2019). Thunderstorm Classification Functions Based on Instability Indices and GNSS IWV for the Sofia Plain.* Remote Sensing, 11*(24). https://doi.org/10.3390/rs11242988.)*

package (TropNET, 2021) developed at Geodetic Observatory Pecny is used (Dousa & Vaclavovic, 2015).

As a part of BeRTISS service, the Bulgarian Integrated NowCAsting tool (BINCA) is developed by HSA. BINCA platform integrates the near real-time GNSS-derived water vapor, NWP products with radar observations. The integrated water vapor from GNSS is derived using the surface pressure and temperature from the WRF model. The WRF model v3.7.1 (WRF, n.d.) is initiated at 00 UTC and computes a 48h forecast for a domain covering Bulgaria. The horizontal resolution is 2km with 45 vertical levels and initial and boundary conditions from the Global Forecasting System model. The GNSS IWV time series and maps, weather radar reflectivity and precipitation as well WRF model animation of temperature, precipitation, and cloud cover are integrated in the BINCA web platform (Bulgarian Integrated NowCAsting Tool (BINCA), 2021). The platform is publicly accessible and has been used in operational hail suppression in Bulgaria since 2020. The BINCA IWV time series for selected station Dolni Dubnik is presented in Fig. 4.20.

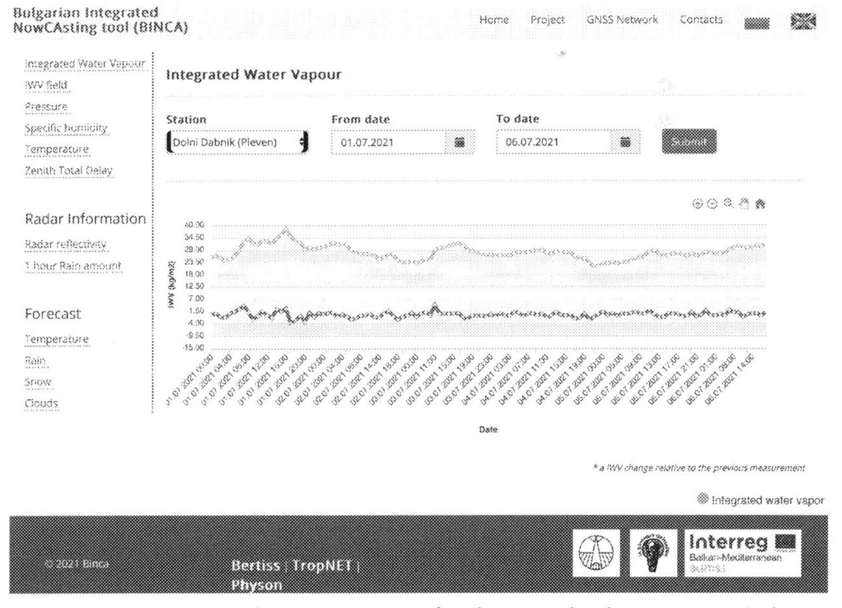

Fig. 4.20 BINCA near real-time IWV series for the period July 1-6, 2021. *(Bulgarian Integrated NowCAsting tool (BINCA). (2021). https://binca-bg.eu/.)*

References

Benevides, P., Catalao, J., & Miranda, P. (2015). On the inclusion of GPS precipitable water vapour in the nowcasting of rainfall. *Natural Hazards and Earth System Sciences, 15*(12), 2605–2616.

Bevis, M., Businger, S., Herring, T. A., Rocken, C., Anthes, R. A., & Ware, R. H. (1992). GPS meteorology: Remote sensing of atmospheric water vapor using the global positioning system. *Journal of Geophysical Research: Atmospheres, 97*(D14), 15787–15801.

Bouma, H. R. (2002). *Ground-based GPS in climate research*. Chalmers University of Technology.

Bulgarian Integrated NowCAsting tool (BINCA). (2021). https://binca-bg.eu/.

Calori, A., Santos, J. R., Blanco, M., Pessano, H., Llamedo, P., Alexander, P., et al. (2016). Ground-based GNSS network and integrated water vapor mapping during the development of severe storms at the Cuyo region (Argentina). *Atmospheric Research, 176*, 267–275.

Davis, J., Herring, T., Shapiro, I., Rogers, A., & Elgered, G. (1985). Geodesy by radio interferometry: Effects of atmospheric modeling errors on estimates of baseline length. *Radio Science, 20*(6), 1593–1607.

Desbois, M., Seze, G., & Szejwach, G. (1982). Automatic classification of clouds on METEOSAT imagery: Application to high-level clouds. *Journal of Applied Meteorology and Climatology, 21*(3), 401–412.

Dousa, J., & Vaclavovic, P. (2015). The evaluation of ground-based GNSS tropospheric products at geodetic observatory Pecný. In *IAG 150 years* (pp. 759–765). Springer.

Elgered, G., Plag, H., van der Marel, P., Barlag, S., & Nash, J. (2005). *COST 716: Exploitation of ground-based GPS for operational numerical weather prediction and climate applications*. Brussels: European Commission.

GNSS Meteorology explained. (n.d.). https://www.youtube.com/watch?v=t1inZaRdWY4.

Guerova, G., Brockmann, E., Quiby, J., Schubiger, F., & Matzler, C. (2003). Validation of NWP mesoscale models with Swiss GPS network AGNES. *Journal of Applied Meteorology, 42*(1), 141–150.

Guerova, G., Brockmann, E., Schubiger, F., Morland, J., & Mätzler, C. (2005). An integrated assessment of measured and modeled integrated water vapor in Switzerland for the period 2001–03. *Journal of Applied Meteorology, 44*(7), 1033–1044.

Guerova, G., Dimitrova, T., & Georgiev, S. (2019). Thunderstorm classification functions based on instability indices and GNSS IWV for the Sofia Plain. *Remote Sensing, 11*-(24). https://doi.org/10.3390/rs11242988.

Haase, J., Ge, M., Vedel, H., & Calais, E. (2003). Accuracy and variability of GPS tropospheric delay measurements of water vapor in the western Mediterranean. *Journal of Applied Meteorology, 42*(11), 1547–1568.

Haralambous, H., Oikonomou, C., Pikridas, C., Guerova, G., Dimitrova, T., Lagouvardos, K., et al. (2018). BeRTISS project: Balkan-Mediterranean real-time severe weather service. In *Vol. 10773. Sixth international conference on remote sensing and geoinformation of the environment (RSCy2018)* (p. 1077313). International Society for Optics and Photonics.

Inoue, H. Y., & Inoue, T. (2007). Characteristics of the water-vapor field over the Kanto district associated with summer thunderstorm activities. *SOLA, 3*, 101–104.

Jones, J., Guerova, G., Douša, J., Dick, G., de Haan, S., Pottiaux, E., et al. (2019). Advanced GNSS tropospheric products for monitoring severe weather events and climate. In *COST action ES1206 final action dissemination report* (p. 563).

Łoś, M., Smolak, K., Guerova, G., & Rohm, W. (2020). GNSS-based machine learning storm Nowcasting. *Remote Sensing, 12*(16). https://doi.org/10.3390/rs12162536.

Ohtani, R., & Naito, I. (2000). Comparisons of GPS-derived precipitable water vapors with radiosonde observations in Japan. *Journal of Geophysical Research: Atmospheres, 105*(D22), 26917–26929.

Schmetz, J., Pili, P., Tjemkes, S., Just, D., Kerkmann, J., Rota, S., et al. (2002). An introduction to Meteosat second generation (MSG). *Bulletin of the American Meteorological Society, 83*(7), 977–992.

Sissenwine, N., Dubin, M., & Wexler, H. (1962). The US standard atmosphere, 1962. *Journal of Geophysical Research, 67*(9), 3627–3630.

TropNET. (2021). Retrieved from https://www.pecny.cz/Joomla25/index.php/trop-net/trop-net-intro. (Accessed 8 February 2021).

Vedel, H., Huang, X., Haase, J., Ge, M., & Calais, E. (2004). Impact of GPS zenith tropospheric delay data on precipitation forecasts in Mediterranean France and Spain. *Geophysical Research Letters, 31*(2).

WRF. (n.d.). https://www.mmm.ucar.edu/weather-research-and-forecasting-model.

Zinner, T., Mannstein, H., & Tafferner, A. (2008). Cb-TRAM: Tracking and monitoring severe convection from onset over rapid development to mature phase using multi-channel Meteosat-8 SEVIRI data. *Meteorology and Atmospheric Physics, 101*(3), 191–210.

CHAPTER 5

GNSS and numerical weather prediction models

Numerical weather prediction

Richardson experiment

In 1922, Lewis Fry Richardson published the first attempt at numerical weather forecasting (Richardson, 2007), applying the ideas developed by W. Bjorknes in his article "The Problem of Weather Forecasting from the Perspective of Mathematics and Mechanics" (1904). In this paper, Bjorknes formulated the time equations with initial conditions for baroclinic fluid (the barocline is the misalignment, or angle, between the isolines of the gradients of pressure and temperature), and Richardson used a hydrostatic approximation to solve them. As a result, the pressure was severely overestimated with a value for the pressure change of 145 hPa for a 6-h interval for Central Europe (Fig. 5.1). The obtained pressure value is much higher than the actual measured pressure changes and the reason for this large error is an imbalance in the initial conditions in the wind and pressure fields. Despite this unsuccessful first attempt, numerical weather forecasting developed very rapidly after the World War II as a result of the development of computers and is now an indispensable part of operational weather forecasting.

Conservation laws of momentum, mass, and heat

To describe mathematically the atmospheric flow, the atmosphere can be considered as an ideal mixture of dry air, water vapor, liquid water, and water in solid state. The atmosphere is subject to the external forces of gravity and the inertial Coriolis force. Internally, various processes due to heat, mass, and momentum transfer as well as phase changes of water may take place. The basic conservation laws for momentum, energy, mass, and heat are then represented by the following equations:

Momentum conservation equation (Navier-Stokes equation):

$$\frac{d}{dt}\mathbf{v} = -2\mathbf{\Omega}\times\mathbf{v} - \frac{1}{\rho}\nabla_3 p + \boldsymbol{g} + \boldsymbol{F} \tag{5.1}$$

Global Navigation Satellite System Monitoring of the Atmosphere
https://doi.org/10.1016/B978-0-12-819152-1.00006-4

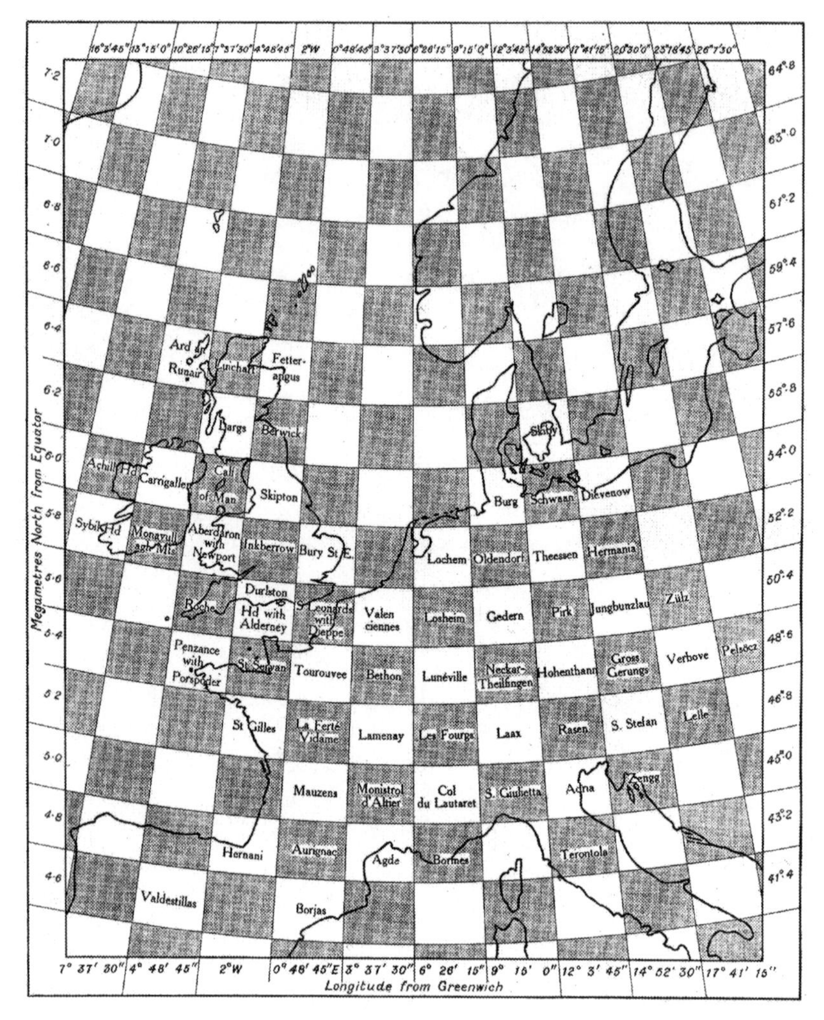

Fig. 5.1 Map of Central Europe with the grid used for Richardson's forecast experiment. *(From Richardson, L. F. (2007). Weather prediction by numerical process. Cambridge University Press.)*

where $\boldsymbol{v}$ is barycentric velocity (relative to the rotating Earth), $\boldsymbol{\Omega}$ is constant angular velocity of Earth's rotation, ρ is total density of the air mixture, p is pressure, $\boldsymbol{g}$ is acceleration of gravity, and $\boldsymbol{F}$ is viscous friction.

Energy/heat conservation equation (first law of thermodynamics):

$$C_v \frac{d}{dt}(\rho q) + p \frac{d}{dt}\left(\frac{1}{\rho}\right) = J \tag{5.2}$$

where C_v is the specific heat on a constant pressure surface, J is the heat flow per unit time per unit mass, and q is the specific humidity.

Mass conservation equation (dry air):

$$\frac{\partial}{\partial t}\rho = -\nabla_3 \cdot (\rho v) \tag{5.3}$$

Mass conservation equation (water vapor):

$$\frac{d}{dt}q = Q_e - Q_c \tag{5.4}$$

where Q_e is water vapor evaporation and Q_c is water vapor condensation.

Ideal gas law (equation of state):

$$p = \rho RT \tag{5.5}$$

where $R = 287.06 \frac{J}{kgK}$ is ideal gas constant.

In April 1950, using the ENIAC computer at the Institute for Advanced Study in Princeton, Jule Charney, Ragnar Fjørtoft, and John von Neumann performed the first weather forecast by solving numerically those equations. This experiment was the basis of the numerical weather prediction (NWP) the way it is known today.

NWP models and data assimilation methods

NWP models can be categorized based on the characteristic times and lengths of the phenomena they represent. The widely used atmospheric-scale classification scheme presented in Fig. 5.2 was proposed by Orlanski (1975). Largest spatiotemporal scale is macroscale with typical horizontal scale 2000–5000 km, time scale from days to a week, and typical phenomena the mid-latitude weather systems of low pressure—cyclone and high pressure—anticyclone. The mesoscale can be defined with spatial range 10–100 km, time range 1–24 h, and typical phenomena thunderstorms, sea breeze, foehn, and fog. The microscale is defined in spatial range below 1 km, temporal range below 1 h, and typical phenomena tornados, dust devils, or turbulence.

NWP model types and setup

Based on the atmospheric scale to be resolved, global and regional NWP models are developed. According to the WMO (Anticipated Advances in Numerical Weather Prediction (NWP), and the Growing Technology

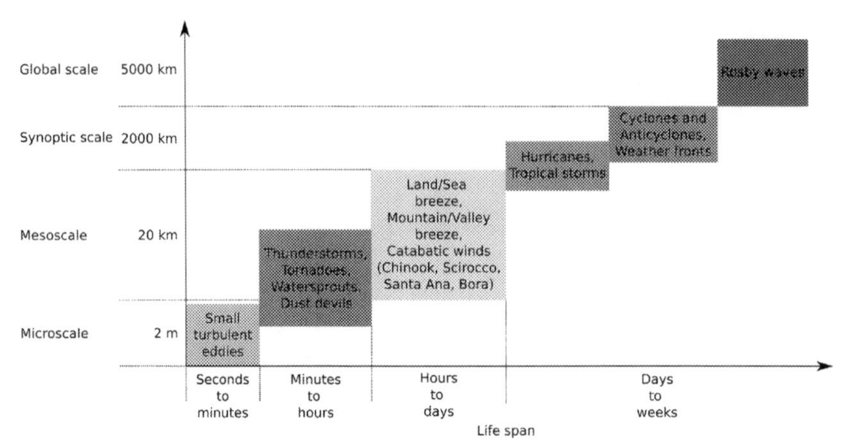

Fig. 5.2 Atmospheric scale classification. *(Reproduced from Orlanski, I. (1975). A rational subdivision of scales for atmospheric processes. Bulletin of the American Meteorological Society, 527–530.)*

Gap in Weather, 2017), the leading NWP centers operate global NWP models with horizontal resolution in the range 16–45 km with majority with resolution below 30 km. For example, ECMWF operates a high-resolution global model (HRES) with a resolution of 9 km and 137 vertical levels. The model mesh and vertical-level distribution is shown in Figs. 5.3 and 5.4. The ECMWF Atmospheric Global Circulation model (2.1 Global Atmospheric Model, n.d.) describes the dynamical evolution of the atmosphere worldwide and is used for medium-range forecasts up to 10 days. It is a general atmospheric model of uniform model physics and structure, which is executed on a global scale. The model uses the most accurate estimate of the current conditions and the most up-to-date description of the model physics and employs throughout modeled land surface conditions (e.g., snow cover, soil moisture), ocean conditions (e.g., sea-surface temperature, sea ice), stratospheric representation, and atmospheric dynamical processes (that together help deliver Rossby wave propagation, weather regime changes, etc.). The regional NWP models have a grid resolution between 1.5 and 12 km and are computed on a limited area. An overview of the operational regional NWP models in Europe is provided via EUMETNET SRNWP program (SRNWP, n.d.). For example, COSMO model (Cosmo, n.d.) is a nonhydrostatic limited-area model designed for both NWP and various scientific applications on the meso-beta and meso-gamma scale. The COSMO model is based on the primitive thermo-hydrodynamical equations describing compressible flow in a moist atmosphere. The model equations are formulated in rotated geographical coordinates and a generalized

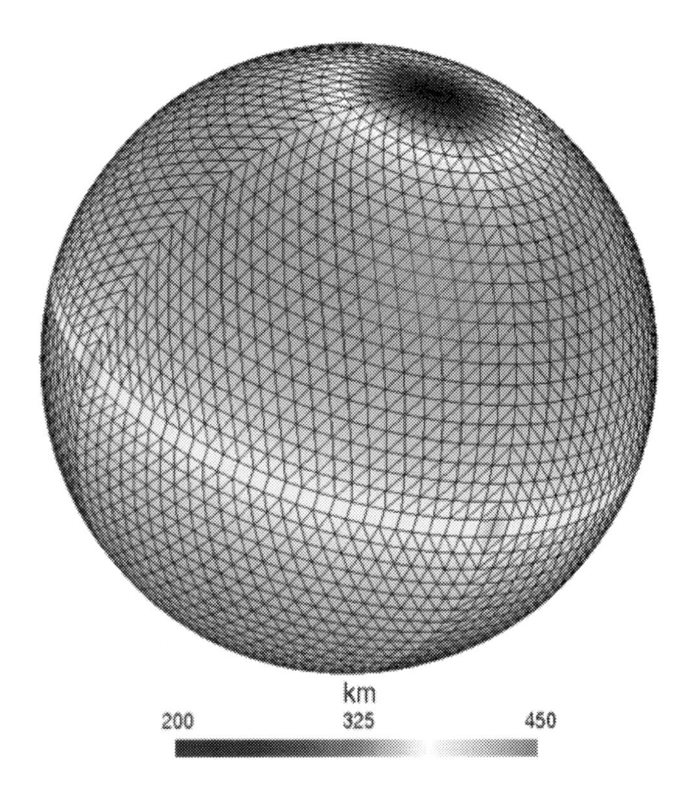

Fig. 5.3 HRES model mesh with horizontal resolution of 9 km. *(Grid point Resolution. (n.d.). Retrieved from https://confluence.ecmwf.int/display/FUG/Grid+point+Resolution. Accessed 5 February 2021.)*

terrain following height coordinate. The regional COSMO-7 model, operated by MeteoSwiss, has a horizontal resolution of 6.6 km, 60 vertical levels, and has a domain over Europe with 393 × 338 grid points. The model produces a 3-day forecast and its initial and boundary conditions are same as that for ECMWF HRES model. MeteoSwiss operates also the COSMO-1 model with horizontal resolution of 1.1 km, 80 vertical levels, and 45-h forecast. COSMO-1 domain covers Switzerland, has a mesh of 1158 × 774 grid points, and uses the ECMWF HRES initial and boundary conditions.

Assimilation of observations in numerical models

The initial state of the atmosphere is of critical importance for accurate weather forecasting. To generate the best possible NWP model initial state, observations of the meteorological elements are used. The developed

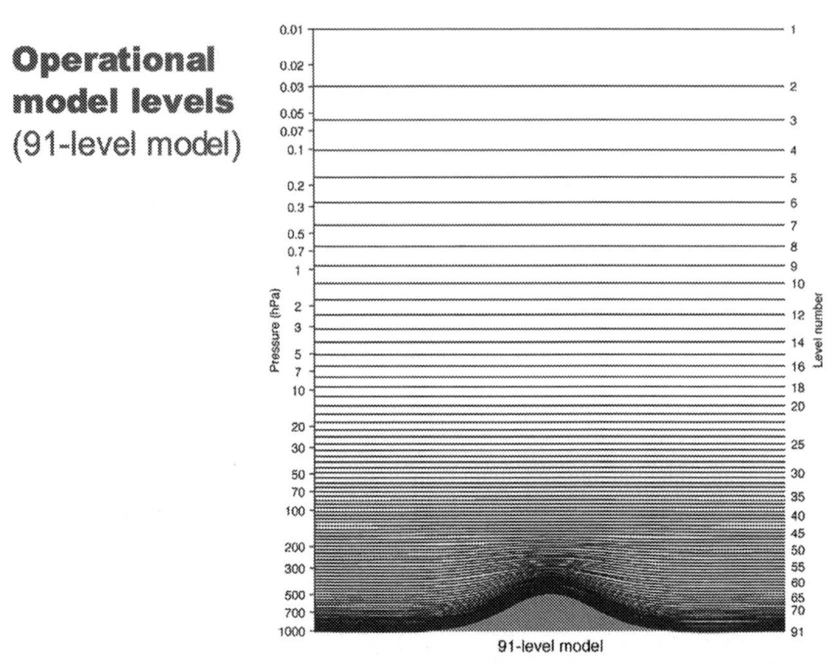

Fig. 5.4 HRES model with vertical levels. *(Grid point Resolution (n.d.). Retrieved from https://confluence.ecmwf.int/display/FUG/Grid+point+Resolution. Accessed 5 February 2021)*

algorithms are named data assimilation. Talagrand (1997) defined data assimilation as "a process in which observations distributed in time are merged together with a dynamical numerical model of the flow in order to determine as accurately as possible the state of the atmosphere." The data assimilation concept is presented in Fig. 5.5. The model prognostic variable derivative in time (left side of Eq. 5.5) is a function of: (1) model dynamics and physics (Q term) and (2) observation increment term, which is the difference between the observed and the initial (background) model state. One of the data assimilation schemes is Newtonian relaxation or nudging (Hoke & Anthes, 1976). The NWP model's prognostic variables are relaxed toward prescribed values within a fixed time window. For example, the prognostic equation for the specific humidity (Guerova, Bettems, Brockmann, & Matzler, 2006) has the following form:

$$\frac{\partial}{\partial t}q = Q(q, u, v, \ldots t) + G_q W_q [q_{obs} - q_{mod}] \tag{5.6}$$

where Q denotes the model physics and dynamics, q_{obs} and q_{mod} are the observed and the model specific humidity, respectively, G_q is the coefficient

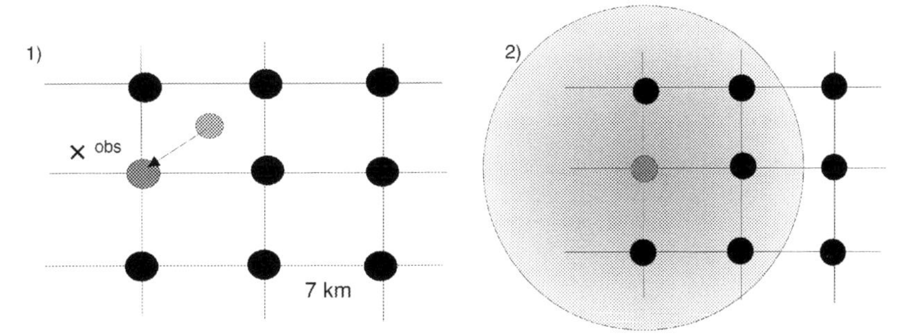

Fig. 5.5 (1) Schematic representation of observation *(blue dot; dark gray in print version)* assimilation in NWP model grid point *(red dot; dark gray in print version)*. (2) Observation spread to the neighborhood grid points within a predefined radius of influence. *(Courtesy: Guergana Guerova.)*

defining the relaxation scale, and W_q consists of spatial and temporal weights and a quality factor. The second term in Eq. (5.5) is the nudging term and $q_{obs} - q_{mod}$ are the observation increments. For the horizontal spreading of the observation increments, an autoregressive weight function is used:

$$W_q(x) = \left(1 - {}^{x}\!/_{s}\right) e^{-x/s} \tag{5.7}$$

where x is the distance between the model grid point and the observation and s is a correlation scale factor. The temporal weight function $W_q\,(t)$ equals one at the observation time and decreases linearly to zero at 1.5 h before and 0.5 h after the observation time, i.e., a 2-h asymmetric sawtooth-shaped time window is used. The observation density is also taken into account to set the value of W_q. The observation increments are computed once every 240 s.

Another widely used assimilation technique is variational assimilation (VAR), which is a form of statistical interpolation and requires an estimation of the error covariances between variables in the background state, as well as error covariances between the observed variables. VAR computes optimal analysis by minimizing a cost function that incorporates the distance between the analysis and observations within the assimilation window. VAR data assimilation requires observation error covariance matrix and background error covariance matrix (Kalnay, 2002; Talagrand, 1997). Three-dimensional variational VAR (3D-Var) scheme uses error covariance matrices from a static climatology, and all observations within a given assimilation window are assumed to be valid at the analysis time. The four-dimensional VAR (4D-Var) scheme minimizes the cost function, subject to the NWP model equations, to find the best model trajectory through the entire assimilation window, rather than just at the analysis time.

GNSS IWV assimilation with nudging technique in the COSMO NWP model

Using the nudging technique, a direct assimilation of GNSS-derived IWV is not possible as no prognostic variable IWV is computed in the NWP model equations. Thus, an indirect assimilation procedure based on a study by Kuo, Guo, and Westwater (1993) has been developed for the COSMO model (Tomassini, Gendt, Dick, Ramatschi, & Schraff, 2002). The steps are shown in Fig. 5.6. First, the zenith total delay (ZTD) from GNSS is converted into IWV (see Chapter 4) using the COSMO NWP model surface temperature and pressure. Second, a K ratio between the observation and the model is calculated. Third, using the K ratio the model specific humidity profile is scaled from surface to the 500-hPa level. Specific humidity is set to saturation at any level, where the specific humidity exceeds its saturation value due to the scaling. The humidity increments are spread laterally using a correlation scale factor of 35 km (as in Eq. 5.6). Only GNSS stations with a difference between station height and model orography smaller than 100 m have

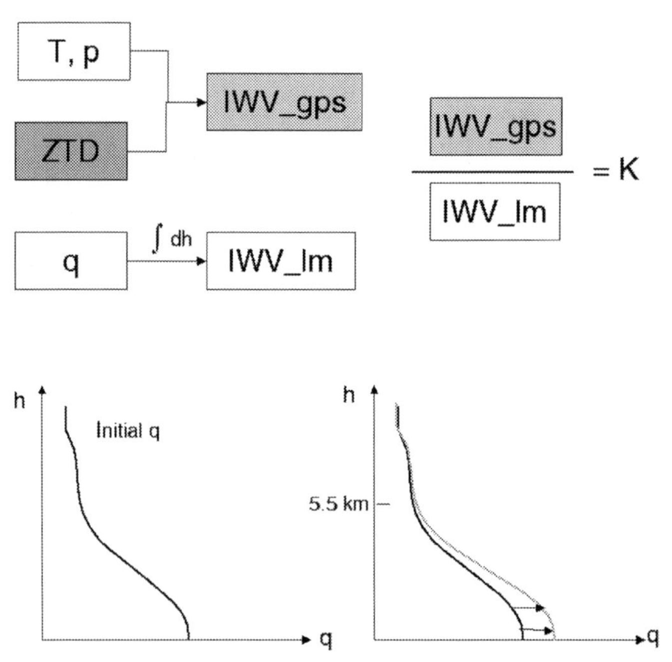

Fig. 5.6 IWV assimilation procedure in the COSMO model with nudging technique. *(From Guerova, G. (2003). Application of GPS derived water vapor for numerical weather prediction in Switzerland. Verlag nicht ermittelbar. https://biblio.unibe.ch/download/eldiss/03guerova_g.pdf.)*

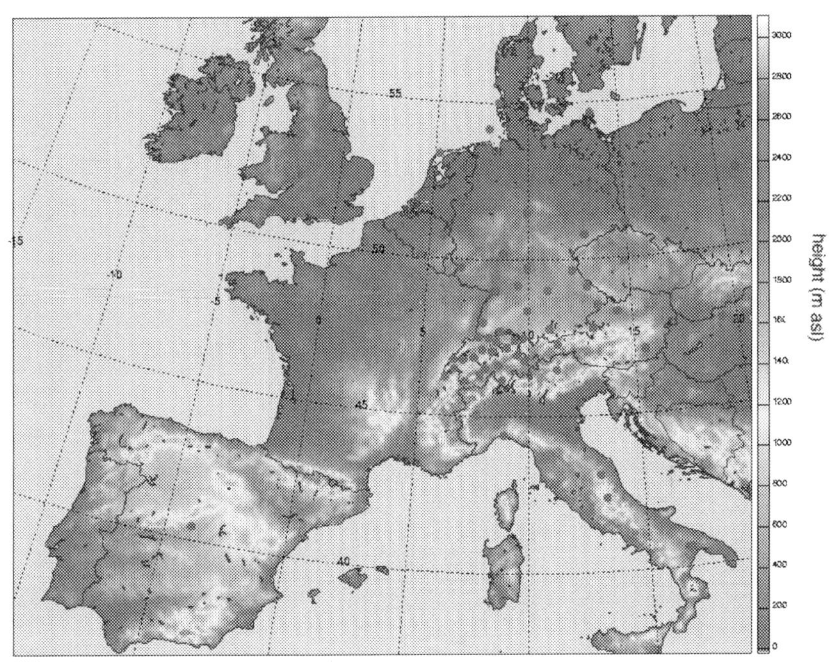

Fig. 5.7 COSMO model domain (color map) with GNSS station assimilated *(red dots; dark gray in print version). (From Guerova, G., Bettems, J.-M., Brockmann, E., & Matzler, Ch. (2004). Assimilation of the GPS-derived integrated water vapor (IWV) in the MeteoSwiss numerical weather prediction model—a first experiment.* Probing the Atmosphere with Geodetic Techniques, *29(2), 177–186. doi:10.1016/j.pce.2004.01.009.)*

been assimilated. For GNSS stations above the model orography, the model level closest to the GNSS height has been used as the initial level in the profile scaling. This condition for the height of the station is particularly important for the Alpine areas in Switzerland, France, Austria, and Italy. Due to this condition, about 20 GNSS stations have not been assimilated, 8 of them from the Swiss GNSS network (Fig. 5.7).

Integrated water vapor

The results of the first assimilation experiments with the MeteoSwiss COSMO model version called aLpine model (aLMo) have been carried out for a 2-week period in September 2001. The selected weeks were characterized by an advective weather regime and heavy precipitation in Switzerland. The model assimilated data from about 80 GNSS stations in Europe (Fig. 5.7). There was a tendency for the IWV to increase during the day hours due to the assimilation of the GNSS data. This increase was significant in the southern part of the model domain (south of the Alps), where the relative IWV change reached 30% compared to the operational model (Fig. 5.8A). The results of the aLMo for Switzerland showed a pronounced positive effect on the 24-h forecast of IWV and precipitation. In one case, GNSS IWV assimilation resulted in a significant increase of model IWV. The reason for this was found to be an inadequate distribution of

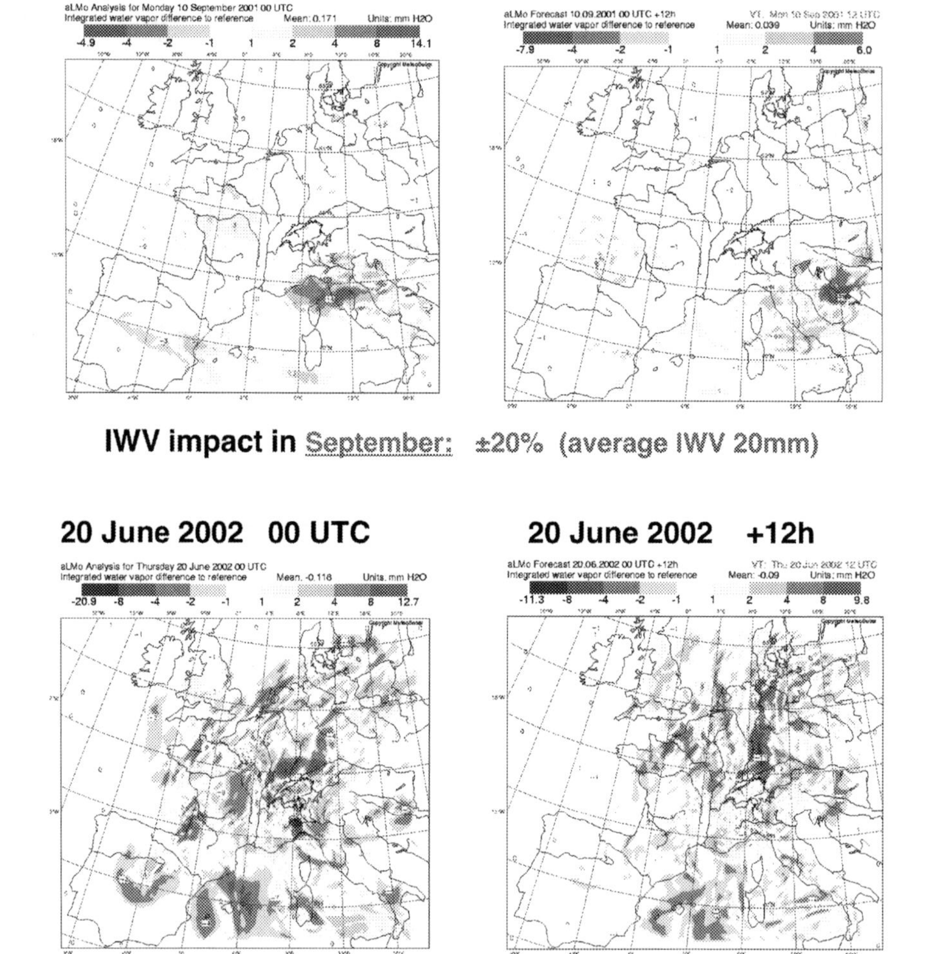

Fig. 5.8 IWV increments on (A) September 10, 2001 at 00 UTC *(left)* and + 12 forecast hour *(right)* and (B) on June 20, 2002 at 00 UTC *(left)* and + 12 forecast hour *(right)*. *(Courtesy: Guergana Guerova.)*

humidity increments. The results from the first GNSS assimilation experiment were encouraging and three new numerical experiments were carried out for different atmospheric circulation regimes. The impact of GNSS IWV on aLMo was found to be significant in June 2002, moderate in September 2001, and insignificant in January 2002.

GNSS assimilation impact on cloud cover and precipitation

GNSS IWV assimilation results in the modification of the vertical and horizontal humidity in the aLMo model. As such, it can be expected that the model hydrology cycle components such as cloud cover and precipitation can be also impacted. This was actually observed in aLMo analysis of cloud cover at 15 UTC on September 9, 2001 at the Gulf of Genoa and inland, south of the Alps. In Fig. 5.9A, it can be seen that IWV assimilation resulted in significant IWV increase in the model by over 20 kg/m^2 [red color (dark gray in print version) in Fig. 5.9A]. As shown in Fig. 5.9B, the observed total cloud cover fraction range is between 40% and 80% over the sea and inland south of the Alps. GNSS IWV assimilation proved beneficial for modeling correctly the cloud cover fraction with values above 80% (Fig. 5.9C). As reported by Guerova, Bettems, Brockmann, and Matzler (2004), the observed cloud cover pattern over the region is better represented in the IWV assimilation experiment.

The GNSS IWV assimilation impact on improvement of NWP model precipitation forecast is highly desirable by the National Meteorological Services. It is well known that NWP models have tendencies to under- or over-predict precipitation due to the complexity of hydrology cycle and phase transition between water vapor and liquid water. Thus, it is logical to expect that assimilation of new observations with high spatiotemporal resolution will correct the model humidity structure, consequently will result in an improved transfer to liquid water phase in model clouds and precipitation. However, due to the complexity of those transfers achieving a positive impact on precipitation by assimilation on GNSS is very difficult. Subjective analysis of GNSS assimilation in aLMo gave mixed results for precipitation impact. Guerova et al. (2004) reported that the positive impact of GNSS IWV on precipitation is: (1) limited to the first 6 h of the model forecast and (2) for cases of heavy precipitation. For example, the GNSS assimilation impact on aLMo model precipitation was reported on June 20, 2002 (Guerova et al., 2006). The weather radar observation of precipitation (Fig. 5.10A) shows two distinct precipitation regions over the northwestern (NW) border of Switzerland and south of the Alps. The intense precipitation

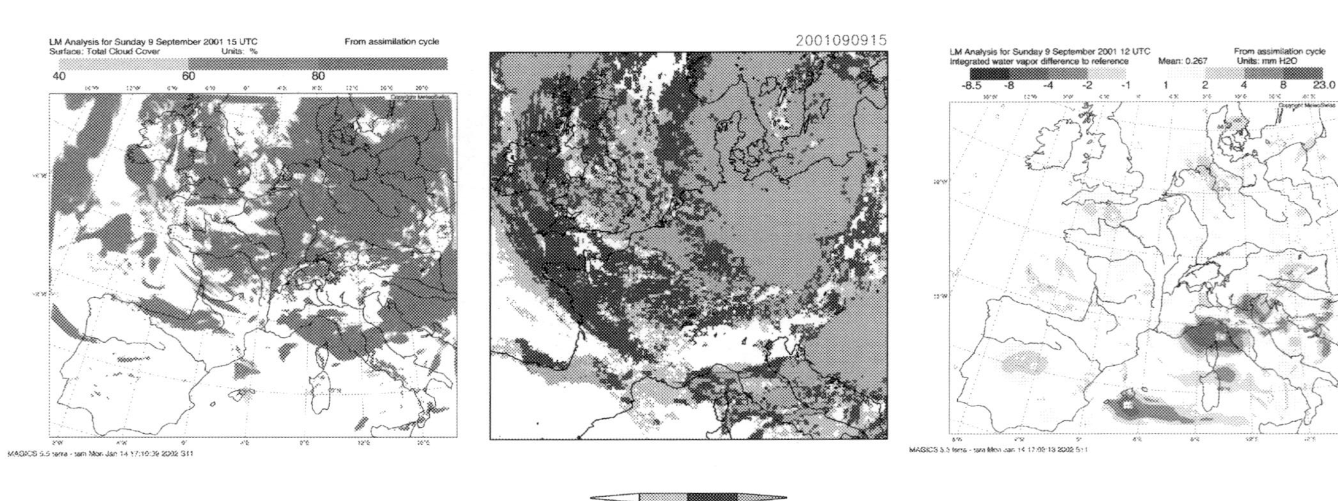

Fig. 5.9 IWV assimilation and cloud cover improvement over the Gulf of Genoa. (A) IWV on at 12 UTC September 9, 2001, (B) observed cloud cover fraction, and (C) aLMo cloud cover fraction after GNSS IWV assimilation. *(From Guerova, G., Bettems, J.-M., Brockmann, E., & Matzler, Ch. (2004). Assimilation of the GPS-derived integrated water vapor (IWV) in the MeteoSwiss numerical weather prediction model—a first experiment.* Probing the Atmosphere with Geodetic Techniques, *29(2), 177–186. doi: 10.1016/j.pce.2004.01.009.)*

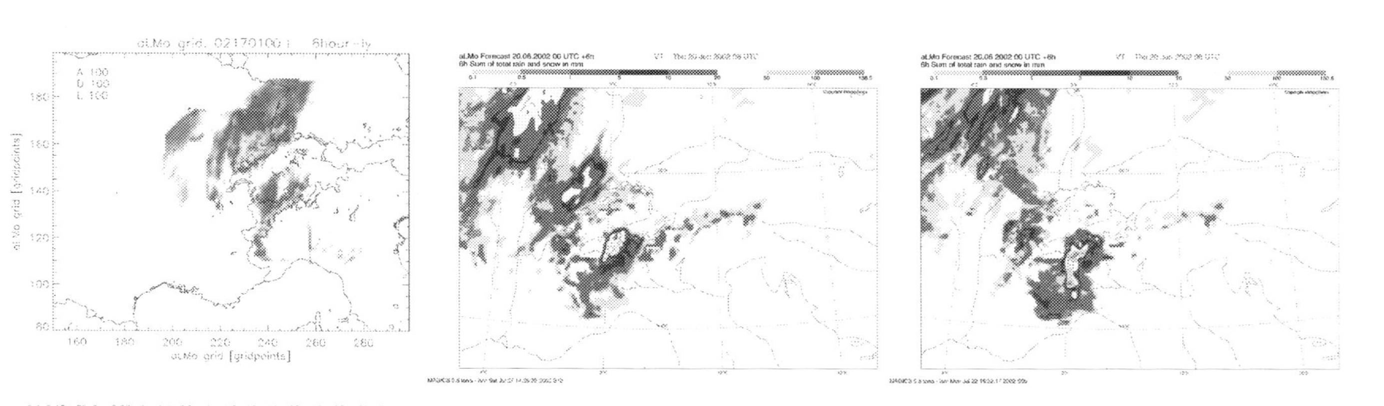

Fig. 5.10 (A) Observed precipitation by weather radar between 00 and 06 UTC on June 20. (B) aLMo reference model precipitation forecast 0 to +6h. (C) aLMo model precipitation forecast after GNSS assimilation on June 20, 2002. *(From Guerova, G., Bettems, J.-M., Brockmann, E., & Matzler, Ch. (2006). Assimilation of COST 716 Near-Real Time GPS data in the nonhydrostatic limited area model used at MeteoSwiss. Meteorology and Atmospheric Physics, 91(1), 149–164. doi: 10.1007/s00703-005-0110-6.)*

over NW Switzerland is well reproduced by the model after GNSS assimilation (Fig. 5.10C) both with respect to intensity and location. As shown in Fig. 5.10B, the reference model precipitation forecast is with much weaker intensity and is clearly displaced to the Southwest Switzerland. From Fig. 5.9, we can see that a moisture deficiency of up to 8 kg/m^2 was present at 00 UTC west of Switzerland in the reference model (the intense orange patch; light gray in print version). The predominant southwesterly flow in the hours between 00 and 06 UTC advected this additional moisture toward Switzerland, which resulted in the reported precipitation prediction improvement. In this case, assimilation of GNSS IWV provides critical correction of the model humidity field ahead of the storm thus benefiting the model forecast of precipitation. This example demonstrated the NWP model sensitivity for spatiotemporal features of the humidity field and the added value of observations from ground-based GNSS networks. This is even more valid today when the observation density in Western Europe increased by factor of 10–20.

GNSS assimilation and diurnal cycle of NWP model

Another advantage of GNSS IWV is the high temporal sampling rate with available hourly to subhourly values. This provides an opportunity for the evaluation of the diurnal cycle of NWP models after assimilation of GNSS observations. One example is presented in Fig. 5.11 for the GNSS station

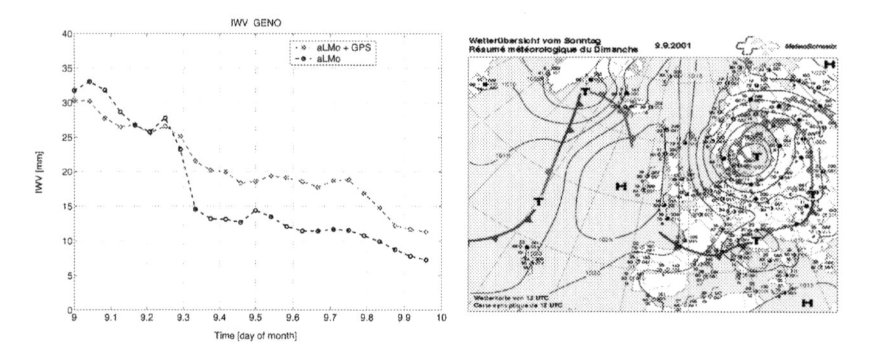

Fig. 5.11 (A) Diurnal IWV cycle for reference model *(dashed black line)* and after GNSS IWV assimilation *(dashed red line; dark gray in print version)* on September 9, 2001 at Genoa Italy. (B) MeteoSwiss mean sea-level pressure analysis map at 12 UTC on September 9, 2001. *(From Guerova, G., Bettems, J.-M., Brockmann, E., & Matzler, Ch. (2004). Assimilation of the GPS-derived integrated water vapor (IWV) in the MeteoSwiss numerical weather prediction model—a first experiment.* Probing the Atmosphere with Geodetic Techniques, *29(2), 177–186. doi: 10.1016/j.pce.2004.01.009.)*

Genoa (GENO). Assimilation of GNSS IWV resulted in a gradual decrease in IWV on September 9, 2001 (dashed red line; dark gray in print version). This was associated with a cold front passage clearly seen on the MeteoSwiss mean sea-level pressure analysis map (Fig. 5.11B). A clear drop of IWV from the reference aLMo run is seen between 08 and 09 UTC. It is interesting to note that both model runs start and end with similar values but during the daytime the difference is substantial. While reported here for 1 day, the model was found to systematically underestimate the daytime IWV and this was linked to the known aLMo tendency to overestimate the light precipitation (Guerova et al., 2004). Another example of IWV assimilation impact on precipitation diurnal cycle is shown in Fig. 5.12. GNSS IWV assimilation results in the improvement of the model diurnal precipitation cycle in particular in the hours between 12 and 16 UTC. Quantitative evaluation of precipitation skill scores (Fig. 5.12B) shows that after GNSS IWV assimilation the precipitation forecast for the thresholds 2 and 10 mm is improved by 10% and 25%, respectively.

GNSS ZTD assimilation with 3D-VAR NWP models of Met Office and Meteo France

GNSS assimilation with the VAR technique allows assimilation of direct observation, namely, ZTD. Due to the inclusion of an error covariance matrix for the NWP model and of physical balances in VAR assimilation, a given observation will influence several model variables simultaneously and in an extended region around the location of the observation. The UK Met Office and Météo France were the first National Meteorological Centers that initiated large-scale operational use of GNSS data in Europe with 3D-VAR and 4D-VAR, following the establishment of E-GVAP in 2005 (Bennitt & Jupp, n.d.; Dousa & Bennitt, 2013; Poli et al., 2007). Today, both institutions report a positive impact from the use of GNSS data in both regional and global NWP models. Fig. 5.13 summarizes the impact per observation assimilated in the Metoffice global NWP model for September 2017. Ground-based GNSS ZTD assimilation has the second largest impact after the drifting buoys observations.

The GNSS ZTD assimilation impact on the Météo France regional model is presented in Fig. 5.14. It can be clearly seen that the surface observations, aircraft profiles, and weather radar have largest impacts on the model. GNSS observations contribute to 3.4% of the information content

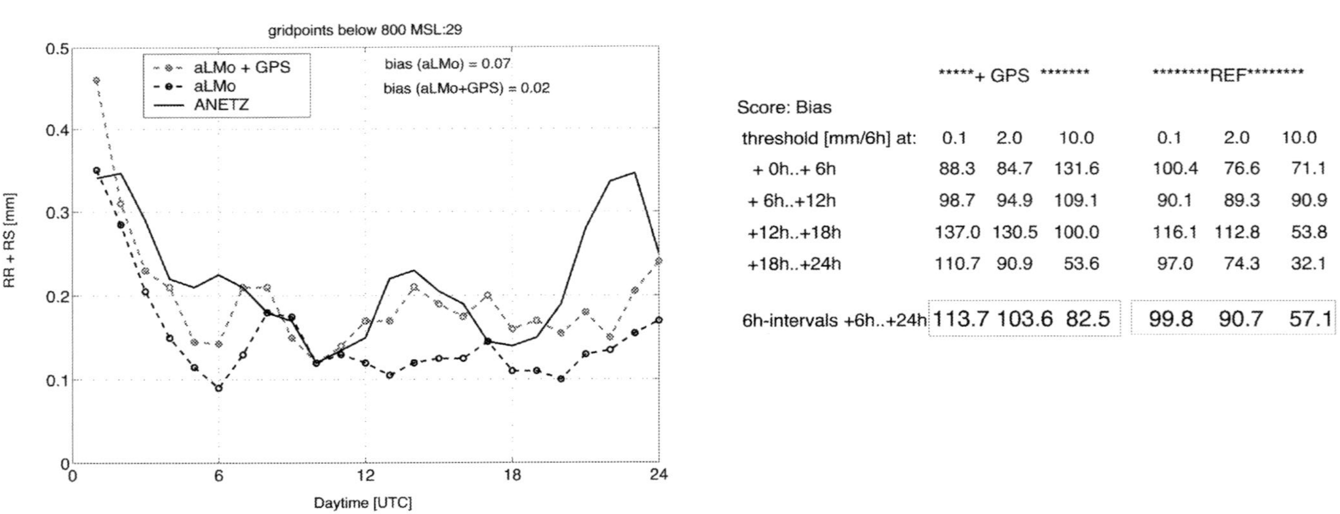

Score: Bias

threshold [mm/6h] at:	*****+ GPS ******* 0.1	2.0	10.0	********REF******** 0.1	2.0	10.0
+ 0h..+ 6h	88.3	84.7	131.6	100.4	76.6	71.1
+ 6h..+12h	98.7	94.9	109.1	90.1	89.3	90.9
+12h..+18h	137.0	130.5	100.0	116.1	112.8	53.8
+18h..+24h	110.7	90.9	53.6	97.0	74.3	32.1
6h-intervals +6h..+24h	113.7	103.6	82.5	99.8	90.7	57.1

Fig. 5.12 (A) Diurnal cycle of precipitation for Switzerland for the period September 9–23, 2001 from synoptic observations *(black line)*, reference model *(dashed black line)*, and after GNSS IWV assimilation *(dashed red line; dark gray in print version)*. (B) Skill score for 6-h accumulated precipitation forecast for thresholds 0.1 mm *(left)*, 2.0 mm *(center)*, and 10.0 mm *(right)*. The best score is 100. *(From Guerova, G., Bettems, J.-M., Brockmann, E., & Matzler, Ch. (2004). Assimilation of the GPS-derived integrated water vapor (IWV) in the MeteoSwiss numerical weather prediction model—a first experiment.* Probing the Atmosphere with Geodetic Techniques*, 29(2), 177–186. doi: 10.1016/j.pce.2004.01.009.)*

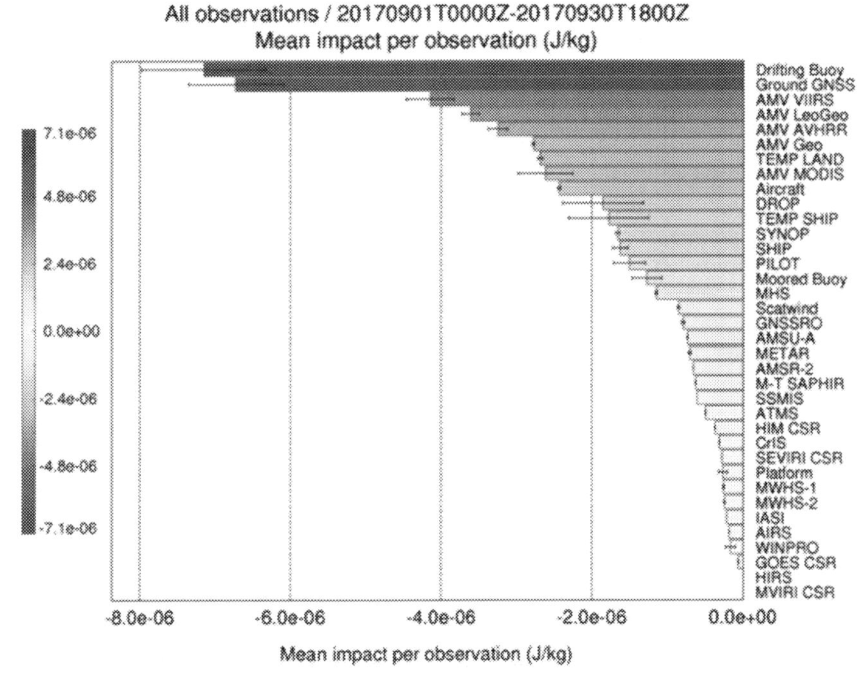

Fig. 5.13 Metoffice global NWP model impact per observation for September 2017. Note that assimilation of ground-based GNSS has the second largest impact. *(From Jones, J., Guerova, G., Douša, J., Dick, G., de Haan, S., Pottiaux, E., & van Malderen, R. (2019). Advanced GNSS tropospheric products for monitoring severe weather events and climate. COST Action ES1206 Final Action Dissemination Report, 563.)*

(pie chart to the left). However, with respect to the information contents of the remote-sensed data (pie chart to the right), GNSS and Meteosat Second Generation (MSG) data are the major contributors.

GNSS STD assimilation in NWP model

The first experiments with assimilation of GNSS slant total delay (SDT) in NWP models were carried out in 2007 but showed mixed impact. Zus et al. (2012) assimilated STD using 4D-VAR technique in MM5 model and reported weak positive impact on precipitation. Kawabata, Shoji, Seko, and Saito (2013) assimilated GNSS STD with 4D-VAR technique during local heavy rainfall event on August 19, 2009 over Okinawa Island, Japan.

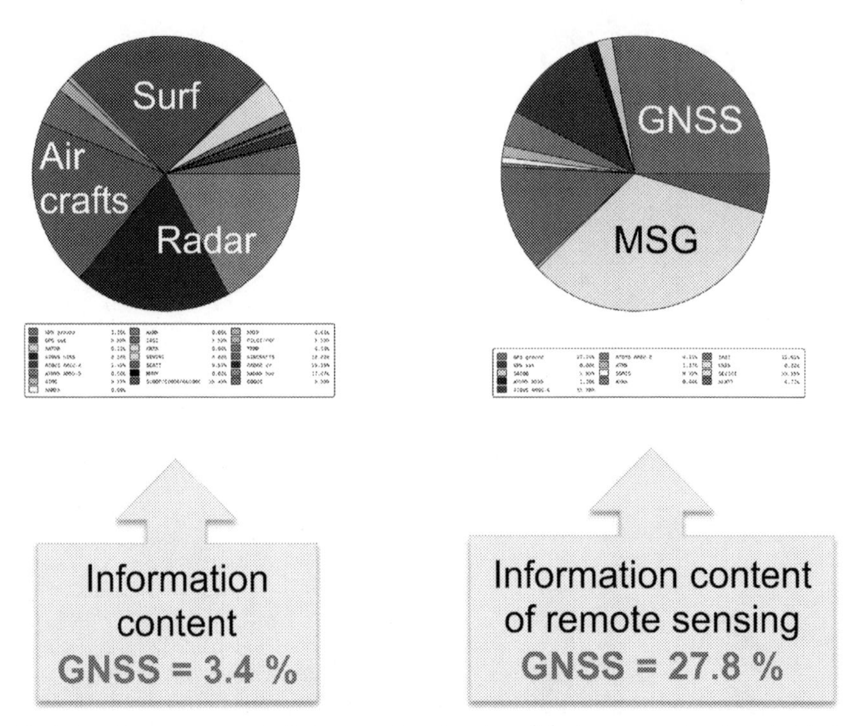

Fig. 5.14 Observation assimilation contribution to Météo France regional model. Left pie chart is for all observations and right for only remote-sensed observations. Red color (*dark gray in print version*) in both charts is for GNSS. *(https://www.isda2014.physik.uni-muenchen.de/program/gpsgnss/isda2014_j_f_mahfouf.pdf.)*

GNSS STD assimilation improved the model water vapor and temperature fields over a wide area and improved precipitation forecast timing and intensity. The vertical cross section of temperature (Fig. 5.15A) shows large positive increments (red color; dark gray in print version) between 500 and 1500 m a.s.l. and large negative increments (blue color; dark gray in print version) between 1000 and 2000 m asl. Water vapor mixing ratio (Fig. 5.15B) and vertical wind speed (Fig. 5.15C) increments increase at the region with enhanced cloud liquid water content (black line). Operational STD assimilation is yet to be considered by the Meteorological Services. However, ongoing preoperational STD assimilation experiments by the German Weather Service (DWD) and a positive impact on precipitation have been reported.

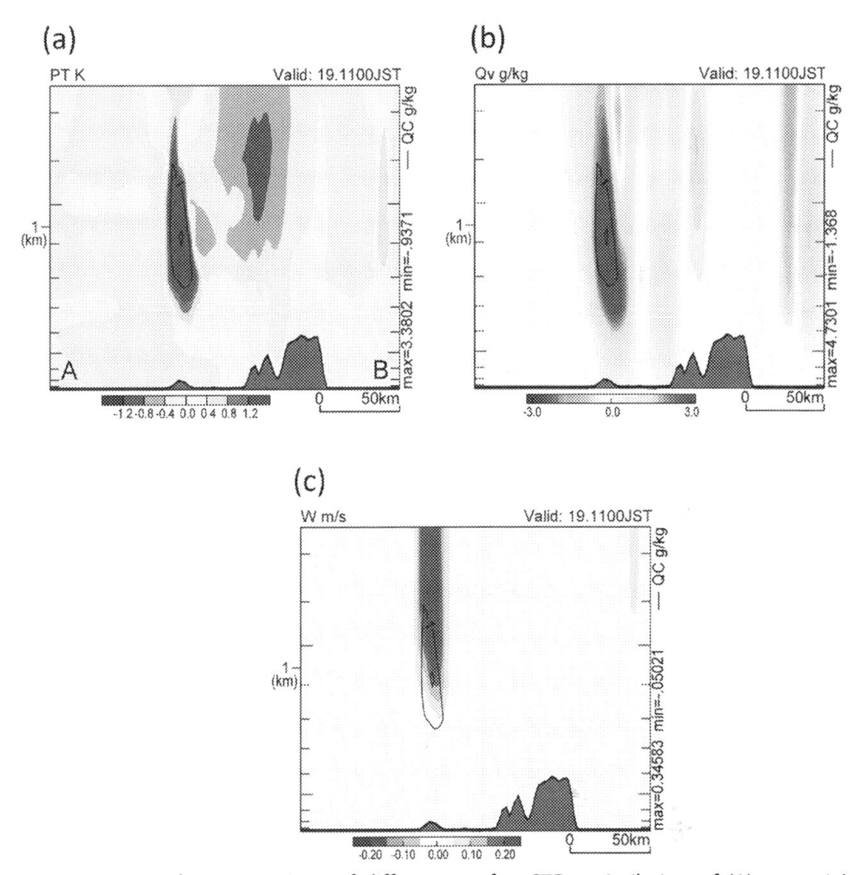

Fig. 5.15 Vertical cross sections of differences after STD assimilation of (A) potential temperature, (B) water vapor mixing ratio, and (C) vertical wind speed. *(From Kawabata, T., Shoji, Y., Seko, H., & Saito, K. (2013). A numerical study on a mesoscale convective system over a subtropical island with 4D-Var assimilation of GPS slant total delays.* Journal of the Meteorological Society of Japan. Ser. II, 91*(5), 705–721.)*

References

2.1 Global Atmospheric Model. (n.d.). https://confluence.ecmwf.int/display/FUG/1.1+Global+Atmospheric+Model.

Anticipated Advances in Numerical Weather Prediction (NWP), and the Growing Technology Gap in Weather. (2017). https://www.wmo.int/pages/prog/www/swfdp/Meetings/documents/Advances_NWP.pdf.

Bennitt, G. V., & Jupp, A. (n.d.). Operational assimilation of GPS zenith total delay observations into the met office numerical weather prediction models. Monthly Weather Review, 140(8), 2706–2719. https://doi.org/10.1175/MWR-D-11-00156.1.

Cosmo NWP. (n.d.). http://www.cosmo-model.org/content/model/default.htm.

Dousa, J., & Bennitt, G. V. (2013). Estimation and evaluation of hourly updated global GPS zenith total delays over ten months. *GPS Solutions, 17*(4), 453–464. https://doi.org/10.1007/s10291-012-0291-7.

Guerova, G., Bettems, J.-M., Brockmann, E., & Matzler, C. (2004). Assimilation of the GPS-derived integrated water vapour (IWV) in the MeteoSwiss numerical weather prediction model—A first experiment. *Probing the Atmosphere with Geodetic Techniques, 29*(2), 177–186. https://doi.org/10.1016/j.pce.2004.01.009.

Guerova, G., Bettems, J.-M., Brockmann, E., & Matzler, C. (2006). Assimilation of COST 716 near-real time GPS data in the nonhydrostatic limited area model used at MeteoSwiss. *Meteorology and Atmospheric Physics, 91*(1), 149–164. https://doi.org/10.1007/s00703-005-0110-6.

Hoke, J. E., & Anthes, R. A. (1976). The initialization of numerical models by a dynamic-initialization technique. *Monthly Weather Review, 104*(12), 1551–1556.

Kalnay, E. (2002). *Atmospheric modeling, data assimilation and predictability*. Cambridge University Press. https://doi.org/10.1017/CBO9780511802270.

Kawabata, T., Shoji, Y., Seko, H., & Saito, K. (2013). A numerical study on a mesoscale convective system over a subtropical island with 4D-Var assimilation of GPS slant total delays. *Journal of the Meteorological Society of Japan Ser II, 91*(5), 705–721.

Kuo, Y.-H., Guo, Y.-R., & Westwater, E. R. (1993). Assimilation of precipitable water measurements into a mesoscale numerical model. *Monthly Weather Review, 121*(4), 1215–1238.

Orlanski, I. (1975). A rational subdivision of scales for atmospheric processes. *Bulletin of the American Meteorological Society, 56*, 527–530.

Poli, P., Moll, P., Rabier, F., Desroziers, G., Chapnik, B., Berre, L., et al. (2007). Forecast impact studies of zenith total delay data from European near real-time GPS stations in Météo France 4DVAR. *Journal of Geophysical Research: Atmospheres, 112*(D6). https://doi.org/10.1029/2006JD007430.

Richardson, L. F. (2007). *Weather prediction by numerical process*. Cambridge university press.

SRNWP. (n.d.). http://srnwp.met.hu.

Talagrand, O. (1997). Assimilation of observations, an introduction (gtSpecial IssueltData assimilation in meteology and oceanography: theory and practice). *Journal of the Meteorological Society of Japan. Ser. II, 75*(1B), 191–209. https://doi.org/10.2151/jmsj1965.75.1B_191.

Tomassini, M., Gendt, G., Dick, G., Ramatschi, M., & Schraff, C. (2002). Monitoring of integrated water vapour from ground-based GPS observations and their assimilation in a limited-area NWP model. *Physics and Chemistry of the Earth, Parts A/B/C, 27*(4–5), 341–346.

Zus, F., Bender, M., Deng, Z., Dick, G., Heise, S., Shang-Guan, M., et al. (2012). A methodology to compute GPS slant total delays in a numerical weather model. *Radio Science, 47*(02), 1–15.

CHAPTER 6

GNSS tomography

History of the tomography method

The tomography method was developed in order to obtain two-dimensional distributions and three-dimensional profiles of the studied objects, by means of integral measurements from different angles and positions. It was proposed in 1930 by radiologist Alessandro Vallebona, and in 1953 (Pollak, 1953) published an article in a medical journal describing the applications of tomography. In medicine, a source and a detector are used to obtain two-dimensional (2D) tomographic images, taking pictures from different angles. Fig. 6.1 shows the source detector system that rotates around the observed object located in the center. The geometry is achieved with observations made from different angles and positions. A necessary condition for obtaining an image is the presence of a large number of observations covering the object. This well-controlled environment allows the reconstruction of high-resolution images. The Nobel Prize in Physiology or Medicine 1979 was awarded jointly to Allan M. Cormack and Godfrey N. Hounsfield "for the development of computer-assisted tomography." The tomography method is widely used in medical practice to examine the human body, with the first photograph of a person with a tomograph taken in 1971. The second application of tomography is seismology (Nolet, 1987). The method has been successfully applied to study the structure of the Earth in order to better understand its structure and composition. The applications of tomography in medicine and geophysics are similar, with a series of flat images being generated in both studies. These images are created by tracing the passage of waves and their energy through the object under study. Two different energy sources can be used in seismology. One is the energy of earthquakes, which generates waves later recorded by receivers on the Earth's surface. With the received information the image of the material is reconstructed in the way of passing of the waves (Fig. 6.2). The other method is generating waves and observing their refraction. This technique is used to obtain information from the object of interest for the study. This is similar to the application in medicine, where information about the object under study is also collected only by the refraction of the waves as

https://doi.org/10.1016/B978-0-12-819152-1.00014-3

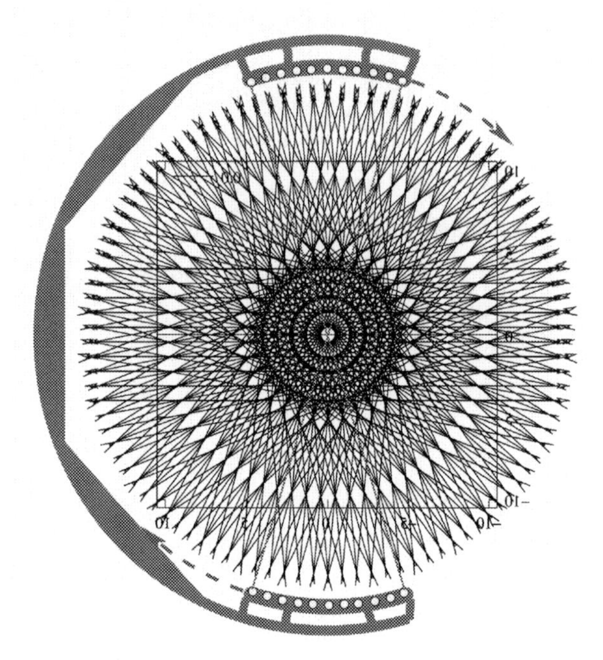

Fig. 6.1 Computer tomography application in the field of medicine. *(https://www.researchgate.net/profile/Galina-Dick/publication/267044071_GNSS_Water_Vapor_Tomography/links/5444b14f0cf2a76a3ccd7afd/GNSS-Water-Vapor-Tomography.pdf.)*

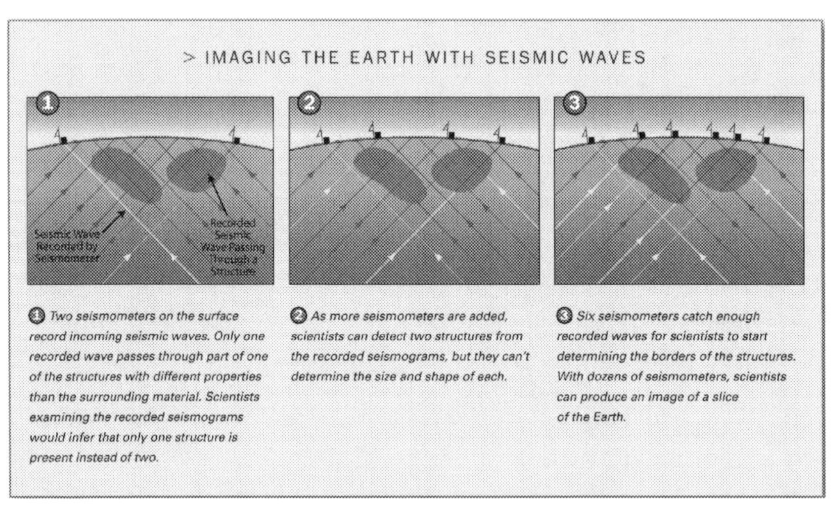

Fig. 6.2 Exemplary scheme of tomography in the field of geophysics. *((N.d.). Retrieved December 2020, 1C.E., from https://www.iris.edu/hq/files/programs/education_and_outreach/lessons_and_resources/docs/es_tomography.pdf.)*

they pass through it. Unlike the application in medicine, the control over the geometry of the source detector in seismology is smaller. However, the advantage is that the field and the detector do not change drastically over time. This time stability allows for network optimization (Gradinarsky, 2002).

Monitoring the troposphere with the tomography method

The troposphere extends from the Earth's surface to a height of 12–18 km and is composed of gases, liquid, and solid particles (aerosols). About 50% of the water vapor in the atmosphere is at an altitude of up to 850 hPa (1.5 km). The use of the tomography method for probing the troposphere was proposed by Flores, Ruffini, and Rius (2000). Fig. 6.3 shows the principle of GNSS tomography for tropospheric sounding. The space above the ground station can be described by a network of the so-called "three-dimensional pixels" or "voxels" (Fig. 6.3). The signal sent by GNSS passes through a large number of voxels and is registered by the ground-based receiver. In each voxel, atmospheric refraction is assumed to be constant. In order to properly apply the tomography method, a network of voxels with a large number of signals passing through it is needed. Ideally, there should be at least one measurement in each voxel on the network. Due to the limited number of satellites and receivers, this is not possible and the network needs to be modified. By modifying the grid, the resolution of the tomography (smaller grid sizes) can be increased at the points where

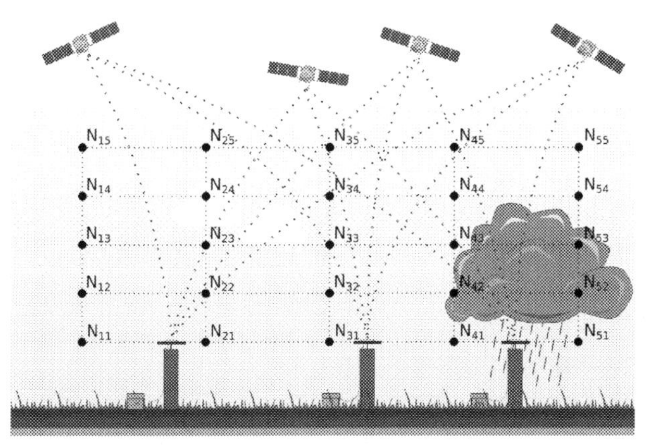

Fig. 6.3 Tomographic network of voxels with a GNSS slanted paths from four satellites to two ground-based receivers. *(Courtesy: Tzvetan Simeonov.)*

more observations intersect, or the resolution (larger grid sizes) can be reduced for the areas with fewer observations. Through observations of GNSS receivers forming a dense local area network, information can be obtained both regarding the amounts of water vapor along the signal path and regarding its three-dimensional structure. The first results using this approach, called GNSS tomography, were successfully applied to water vapor refraction. For operational weather forecasting it is necessary to accurately determine the distribution of water vapor in the atmosphere and its change over time. The temporal and spatial information about the distribution of water vapor, which is obtained by the GNSS tomography method, is of great interest. Several models have been developed for the realization of tomography:

Local Tropospheric Tomography Software—LOTTOS (Flores et al., 2000) uses GNSS data. Simulations and comparisons between the tomography method and real data have been made. In Japan (Hirahara, 2000; Seko, Nakamura, Shoji, & Iwabuchi, 2004) developed a tomographic software package with the main goal of studying water vapor during the Asian monsoons. In Switzerland, (Kruse, 2000) developed AWATOS (Atmospheric Water Vapor Tomography) software and (Troller, Bürki, Cocard, Geiger, & Kahle, 2002) performed numerical experiments and analysis of the obtained results. The method developed by Gradinarsky (2002) is based on the use of a Kalman filter.

Tomographic equation

Atmospheric refraction

In this chapter we present tomographic equations following the approach used by Gradinarsky (2002). As the refractive index of the atmosphere (n) is not equal to one (due to the presence of gases) for electromagnetic waves in the radio frequency range, they propagate with a delay $\Delta L\,(\epsilon, \phi)$ compared to propagation in vacuum. The longer the path through the atmosphere, the greater the delay. The delay varies mainly with altitude ε from the direction of signal arrival, but also with azimuth φ for inhomogeneous atmosphere. The delay can be represented as

$$\Delta L(\epsilon, \phi) = \int_S n\, dx - G \tag{6.1}$$

In Eq. (6.1) the first term specifies the shortest path of radio signal propagation and the second term G is the shortest geometric path. We can represent the above equation as

$$\Delta L(\epsilon, \phi) = \int_S (n-1)dx + S - G \tag{6.2}$$

where S is the geometric path of the refracted signal $S = \int_S dx$. The first term in Eq. (6.2) is the delay of the signal due to the reduced propagation speed due to the refractive index. The difference between the second and the third term takes into account the geometric delay caused by the signal distortion. The geometric delay can be neglected for viewing angles above 15 degree, but should be taken into account for smaller angles (5 degree), where it reaches values of 10 cm. For convenience, the refractive index n is represented as refraction N:

$$N = 10^6(n-1) \tag{6.3}$$

Thus Eq. (6.2) is modified to

$$\Delta L(\epsilon\phi) = 10^{-6}\int_S N(s)dx \tag{6.4}$$

Hydrostatic and wet delays

Due to the different nature of the gases that make up the atmosphere, the delay of the GNSS signal can be divided into hydrostatic and wet. The wet delay is due to the induced and constant dipole moment of water vapor molecules. The hydrostatic delay is due to the induced dipole moment of all other gases in the atmosphere. Both effects are nondispersive at microwave frequencies. Using the refraction representation for a gas composed of q components (Debye, 1929) and transforming it for the case of the atmosphere, we obtain

$$N_{atm} = \sum_{i=1}^{q(dry)} A_i\rho_i + \left(A_V + \frac{B_v}{T}\right)\rho_v \tag{6.5}$$

where the first term represents the dry delay and the second and third terms take into account the distribution of water vapor. The constants A_i and B_i describe the induction and orientation, respectively, of the polarization of the molecules. B_i is zero for dry gases, T is the temperature, and ρ_i is the density of the ith component. Using the equation of state of the ith component, $p_i = Z_i\, \rho_i\, R_i\, T$ (R_i is the gas constant) (Elgered, 1993), the refraction N of the neutral atmosphere is written as

$$N_{atm} = k_1\frac{p_d}{T}Z_d^{-1} + k_2\frac{p_\omega}{T}Z_\omega^{-1} + k_3\frac{p_\omega}{T^2}Z_\omega^{-1} \tag{6.6}$$

where p_d is the partial pressure of the dry component, p_ω is the partial pressure of water vapor, and Z_d^{-1} and Z_ω^{-1} are the reciprocal values of the compressibility of dry and wet air. The values of the constants k and the compressibility of dry and wet air are given by Davis, Herring, Shapiro, Rogers, and Elgered (1985). It is possible to represent the first term of Eq. (6.6) using the total density $\rho=\rho_\omega+\rho_d$ rather than the densities of the wet and hydrostatic components (Davis et al., 1985).

$$N_{atm} = k_1 R_d \rho + k_2' \frac{p_\omega}{T} Z_\omega^{-1} + k_3 \frac{p_\omega}{T^2} Z_\omega^{-1} \tag{6.7}$$

where $k_2' = k_2 - k_1[M_\omega/M_d]$, and M_ω and M_d are the molar masses of water vapor and dry air, respectively, and R_d is the specific gas constant for dry air ($R_d = R/M_d$). This form of the first term of refraction allows integration into Eq. (6.4) without the knowledge of the profiles of different heights of the dry components (Davis et al., 1985). This dry delay, derived from the hydrostatic equilibrium, is henceforth referred to as hydrostatic delay:

$$\Delta L_h = \left(2.2768\,10^{-3} \pm 5\,10^{-7}\right) \frac{P_0}{f(\lambda, H)} \tag{6.8}$$

$$f(\lambda, H) = 1 - 2.66\,10^{-3} \cos 2\lambda - 2.8\,10^{-7} H \tag{6.9}$$

where ΔL_h is in meters, P_0 is the surface pressure in hPa, and λ and H are the latitude in degrees and the height above the geoid in meters, respectively. The other two terms are related to water vapor:

$$N_\omega = k_2' \frac{p_\omega}{T} Z_\omega^{-1} + k_3' \frac{p_\omega}{T^2} Z_\omega^{-1} \tag{6.10}$$

In Eq. (6.10), the effect of compressibility can be neglected (less than the error of the whole formula) (Rüeger, 2002). The total amount of water vapor can be obtained by integrating through the signal propagation path. As obtaining water vapor profile only through ground-based measurements of temperature, pressure, and humidity is difficult, the use of additional observations such as the use of a microwave radiometer or other techniques is necessary. Often, temperature, pressure, and relative humidity profiles obtained from aerological sounding are used to calculate the integrated water delay (ΔL_ω). The delay can be computed from:

$$\Delta L_\omega = 10^{-6} k_3' \int \frac{p_\omega}{T^2}\, ds \tag{6.11}$$

where $k_3' = k_3 + k_2' T_m$ and T_m is the mean temperature of the atmosphere given by

$$\int \frac{e}{T}\,dz = T_m \int \frac{p_\omega}{T^2}\,dz \tag{6.12}$$

To calculate the delay, the average atmospheric temperature for a particular location must be determined. This can be achieved either by integrating an ensemble of aerological sounding profiles and finding values for each season and location, or by assuming a linear relationship between ground temperatures and "average temperature" values. These coefficients are found by using a fit with linear regression to establish the values. If appropriate constants k are used, the expected error is less than 0.2 mm for the zenith delay (Davis et al., 1985).

Slant wet delay

The tomography method based on slanted total delay (STD) (Ware, Alber, Rocken, & Solheim, 1997) has the following advantages: (1) a larger set of observations is used, (2) the observations contain more information about the inhomogeneous atmosphere gradients of temperature and humidity, while ZTD (zenith total delay) is used under a homogeneous atmosphere, and (3) ZTD can be considered as a specially symmetrical case of spatially distributed STDs. STD provides information on the amount of water vapor using the tomography method, which can be used to make 3D profiles of its distribution in the atmosphere.

STDs are calculated with the following equation:

$$STD = m_h ZHD + m_w (ZWD + \cot \epsilon \,(G_n \cos \phi + G_e \sin \phi)) + \delta \tag{6.13}$$

where ZHD and ZWD are the zenith hydrostatic and wet delays, respectively, m_h and m_w are the hydrostatic and wet mapping functions, respectively, for projection to the zenith, G_n and G_e are the gradients of the delays in the north and east direction, respectively, ε is the angle, ϕ is the latitude, and δ is the residual. In order to calculate the slanted delay, it is necessary to distinguish ZHD from ZWD. After using Eqs. (6.4), (6.7), which give the two zenith delays and substituting in Eq. (6.13), the slanted delays can be obtained.

GNSS tomography equation system

The task of the tomography method is to compile and solve a system of equations. In tropospheric tomography, measurements are possible only in a certain geometric configuration, depending on the location of the satellite and the ground receiver. The number of GNSS signals passing through each network voxel depends on the density of the voxel network. The network, in turn, depends on the number of layers and the number of voxels in one layer, the distribution of ground stations, the position of the satellites, and the observation time. In practice, there are not enough satellites and ground stations to give measurements in each voxel. In some voxels, there are many measurements, while in others there are none at all. The whole system of equations is not well defined and there is a problem with its unambiguous solution, which is the main problem of the tomography method.

To find the sloping delay of the GNSS signal in the atmosphere, the following equation is used:

$$STD = 10^{-6} \int_{s} N \, ds \tag{6.14}$$

where N is the refraction of the atmosphere and S is the GNSS signal pathway. In this case, the refractive field N must be found and reconstructed. To solve Eq. (6.14), a network of voxels is constructed in which refraction is assumed to be a constant. The following equation is obtained:

$$A \cdot x = m \tag{6.15}$$

where m is the vector representing the observations or the sloping water delays, x gives the instantaneous state of the atmospheric refraction N_j in each voxel j and the matrix A gives the state x of the observation m.

The system of equations for the simplified model of four satellites and two terrestrial receivers presented in Fig. 6.3 are as follows:

$$l_{11}N_1 + 0\,N_2 + 0\,N_3 + l_{14}N_4 + 0\,N_5 + 0\,N_6 = STD_1 \tag{6.16}$$

$$0\,N_1 + l_{22}N_2 + l_{23}N_3 + l_{24}N_4 + l_{25}N_5 + 0\,N_6 = STD_2 \tag{6.17}$$

$$0\,N_1 + l_{32}N_2 + 0\,N_3 + 0\,N_4 + l_{35}N_5 + l_{36}N_6 = STD_3 \tag{6.18}$$

$$0\,N_1 + 0\,N_2 + l_{43}N_3 + 0\,N_4 + 0\,N_5 + l_{46}N_6 = STD_4 \tag{6.19}$$

In this system, the knowns are the slant delays STD_1, STD_2, STD_3, and STD_4 and the unknowns are the refractions for each voxel N_1, N_2, N_3, N_4, N_5, and N_6 and they are determined by solving the system. Different methods for solving the system have been developed, all of which depend on the matrix of equations. The matrix depends only on the network and the geometry of the observations and the chosen model of the voxels. In order to uniquely determine the matrix, it must meet the following conditions:

At least one GNSS signal must pass through each voxel. Thus, there is no problem with solving the system of equations. This is not possible, but the application of tomography for appropriately determined heights of GNSS stations leads to improved solutions.

The path of the zenith delays is not sufficient to compile an appropriate system of equations, no matter how many. With sloping water delays, it is possible to determine a voxel, even if there is no GNSS receiver in it. A large number of sloping water delays are needed, at different angles, to improve the solution.

The quality of the measured sloping delays limits the sensitivity of the solution. Therefore, it is necessary to make restrictions in order to reduce the influence of noise caused most often by repeated reflections of the GNSS signal.

GNSS tomography—Application

Tomographic reconstruction for Germany

Fig. 6.4A shows the GNSS stations in Germany, which provide tilt delay data for a tomographic model. To cover the area of $700 \times 900 \times 10\,\text{km}^3$, a network with a horizontal resolution of 35–40 km and a vertical resolution of 200–500 m has been built (Bender et al., 2011). Measurements are taken every 30 min, and the total number of voxels is between 10,000 and 15,000. Fig. 6.4B shows the sloping delay paths of the different GNSS systems.

Reconstruction of the water vapor field distribution for July 14, 2009 is shown in Fig. 6.5. A total of 31,222 slant delays were used during the period 22:00–23:00 UTC by GNSS alone. The figure shows cold and dry (blue; dark gray in print version) and warm and humid (red; dark gray in print version) air masses. The left panel shows the water vapor field obtained from a mesoscale numerical weather forecast model (COSMO-DE), the middle panel shows the tomographic reconstruction, and the right panel shows the integrated water vapor field obtained by the GNSS meteorology method (Chapter 4). GNSS stations are represented by black dots. Separate meridional isolines can be seen in the IWV map. At 22:00 UTC, the water vapor field of the numerical model shows high values and gradients concentrated in the southwestern part of the region, while the northern part is dry. According to COSMO-DE, the convergence line is heading southwest. Tomographic reconstruction gives a water vapor field close to that of the GNSS meteorology method and the COSMO-DE model. The tomographic reconstruction of the water vapor field is close to that of the COSMO-DE model, since the COSMO-DE field at 18:00 UTC is used to initialize the tomographic reconstruction. The isolated structures in the

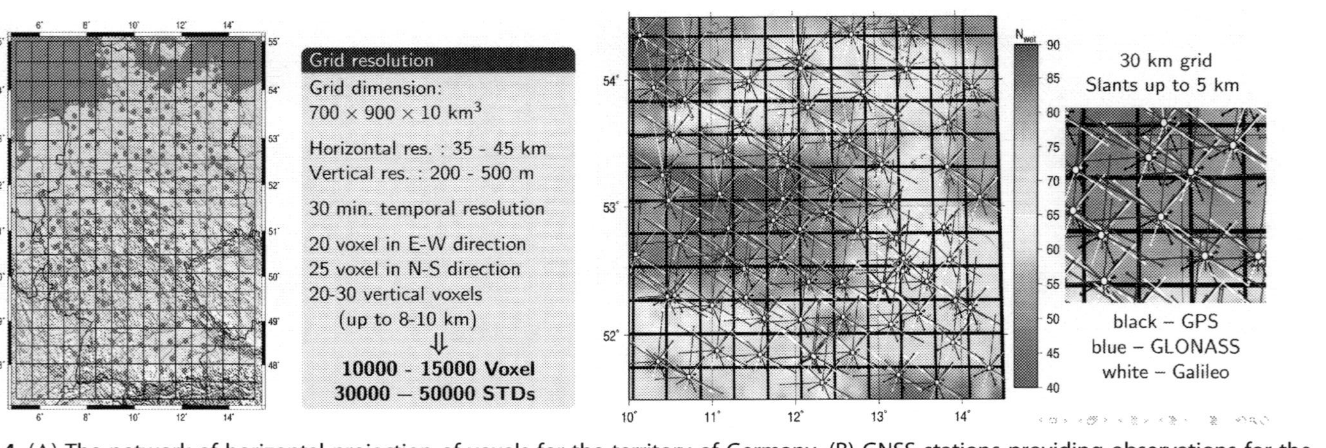

Fig. 6.4 (A) The network of horizontal projection of voxels for the territory of Germany. (B) GNSS stations providing observations for the tomography method are marked with *red dots* (dark gray in print version). *((N.d.). Retrieved from https://www.isda2014.physik.uni-muenchen.de/program/gpsgnss/isda2014_michael_bender.pdf. Accessed 1 December 2020.)*

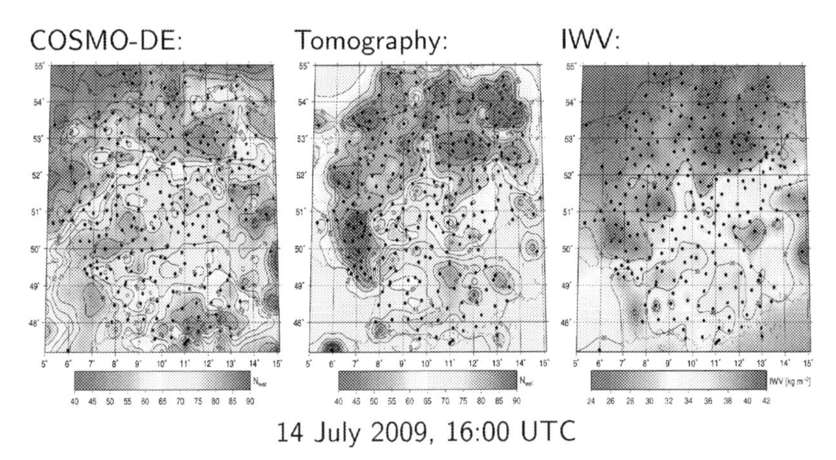

Fig. 6.5 Comparison of GNSS tomography *(middle panel)* with the water vapor field of the numerical model COSMO-DE *(left panel)*, and the field of integrated water vapor *(right panel). ((N.d.). Retrieved from https://www.isda2014.physik.uni-muenchen.de/program/gpsgnss/isda2014_michael_bender.pdf. Accessed 1 December 2020.)*

reconstructed field are mainly due to the insufficient number of stations. The distance between them is 30–60 km, which leads to fewer observations and missing ones in some voxels. The shape of the convergent line in the tomographic reconstruction is closer to that obtained by the GNSS method meteorology as the difference is in the drier southern part. This is due to the density of the GNSS network in this part of Germany, which is not sufficient to determine the sloping water delays. However, the water vapor field reconstructed by the tomography method is better than that of the numerical weather forecast model COSMO-DE. The vertical sections of the water vapor refraction field of the numerical model COSMO-DE (upper panel of Fig. 6.6) and that of the tomographic reconstruction (lower panel of Fig. 6.6) show that the relatively dry area in the northern part of the model has been replaced by a much more wet area in tomographic reconstruction. This is probably due to the fact that the humid air mass in this area is located at a higher altitude.

Tomographic reconstruction of severe weather events

Brenot et al. (2014) presented a case study of GNSS tomographic reconstruction with temporal resolution of 15 min, horizontal resolution of 5 km, and vertical resolution of 500 m. They reported that the west-to-east humidity decrease is associated with warm front passage and precipitation on

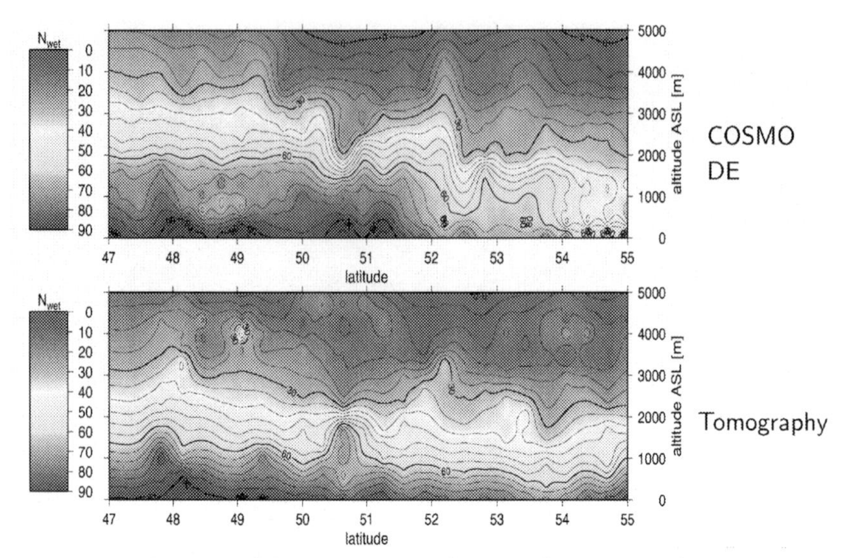

Fig. 6.6 Vertical section of the atmospheric refraction of water vapor for July 14, 2009. The upper panel is from the numerical model COSMO-DE, and the lower from the tomographic reconstruction. *((N.d.). Retrieved from https://www.isda2014.physik.uni-muenchen.de/program/gpsgnss/isda2014_michael_bender.pdf. Accessed 1 December 2020.)*

October 21, 2002. In particular, the water vapor density at 20:00 UTC (Fig. 6.7, c5) was found to decrease in the lower 500 m layer close to the terrain in the NW part of the domain and then decreased in the NNW–SSE part of the domain. In addition, the vertical cross section of water vapor density at 20:00 UTC (Fig. 6.7, d5) shows a marked depletion in the west part of the domain with $6\,g/m^3$ decreasing from 1700 m a.s.l. at 19:45 to 1200 m a.s.l. at 20:15. In contrast, the weather radar reflectivity (Fig. 6.7, b4–b5) increased with water vapor depletion, which is likely the result of condensation and transfer into liquid water. The water vapor depletion is also seen in the IWV map (Fig. 6.7, a4).

Another example of tomographic reconstruction of wet refractivity during super cell thunderstorms in Australia is presented by Zhang et al. (2015). The GNSS wet refractivity profiles shown in Fig. 6.8A are with temporal resolution of 10 min for a 10-day period. From day of year (*doy*) 63 to 64 a clear increase in wet refractivity can be seen (Fig. 6.8A), which correlates with linear increase in radar reflectivity (black line; dark gray in print version in Fig. 6.8B). After the passage of a cold front on the *doy* 64 wet refractivity build-up continued until the heavy rainfall, hail, and flash flooding reached the city of Melbourne on *doy* 65 (blue line; dark gray in print version).

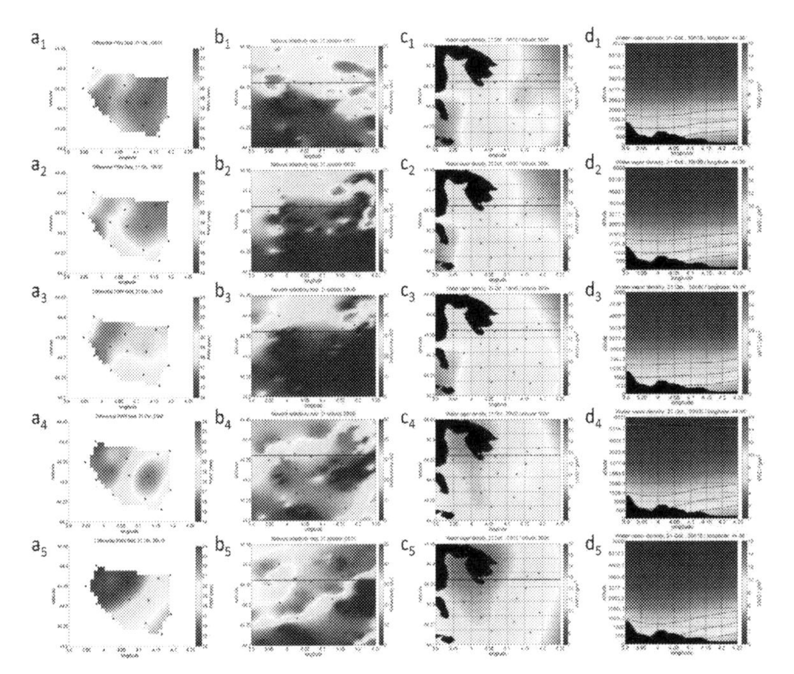

Fig. 6.7 (A) Differential IWV on October 21, 2002 from 19:15 to 20:15, (B) 2D weather radar reflectivity fields [dbZ], (C) GNSS tomography water vapor density (g/m^3) at 500 m asl, and (D) water vapor density vertical cross sections. The *dashed line* in (B) and (C) indicates the location of the vertical cross section shown in (D). *(From Brenot, H., Walpersdorf, A., Reverdy, M., Van Baelen, J., Ducrocq, V., & Champollion, C., et al. (2014). A GPS network for tropospheric tomography in the framework of the Mediterranean hydrometeorological observatory Cévennes-Vivarais (southeastern France). Atmospheric Measurement Techniques, 7(2), 553–578. https://doi.org/10.5194/amt-7-553-2014.)*

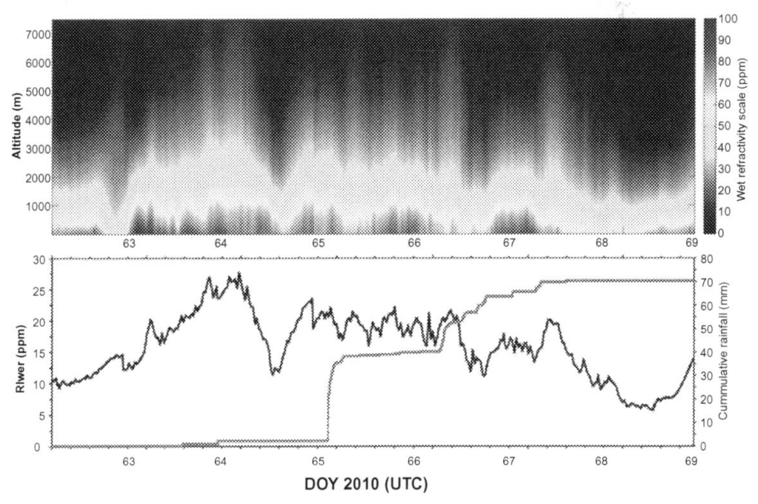

Fig. 6.8 (A) Vertical cross section of wet refractivity from doy 62 to 69 (March 3–8, 2010) and (B) observed radar reflectivity *(black line)* and accumulated precipitation *(blue line; dark gray in print version). (From Zhang, K., Manning, T., Wu, S., Rohm, W., Silcock, D., & Choy, S. (2015). Capturing the signature of severe weather events in Australia using GPS measurements.* IEEE Journal of Selected Topics in Applied Earth Observations and Remote Sensing, 8*(4), 1839–1847.)*

The spatiotemporal evolution of wet refractivity during the storm development is presented in Fig. 6.9 as well as the wet refractivity profile and interpolated radar refractivity from 22:00 UTC on 5 March to 06:00 UTC on 6 March. The evolution of the wet refractivity shows a clear decrease in all layers 4h ahead of the storm at 5:40 UTC. Well visible at 03:00 and 04:00 UTC is the large increase in wet refractivity in the SE part of the region. The large dry-moist gradient seen before the storm at 03:00 and 04:00 UTC indicated strong instability, which predisposes the strong updraft and formation of a mature convective cell.

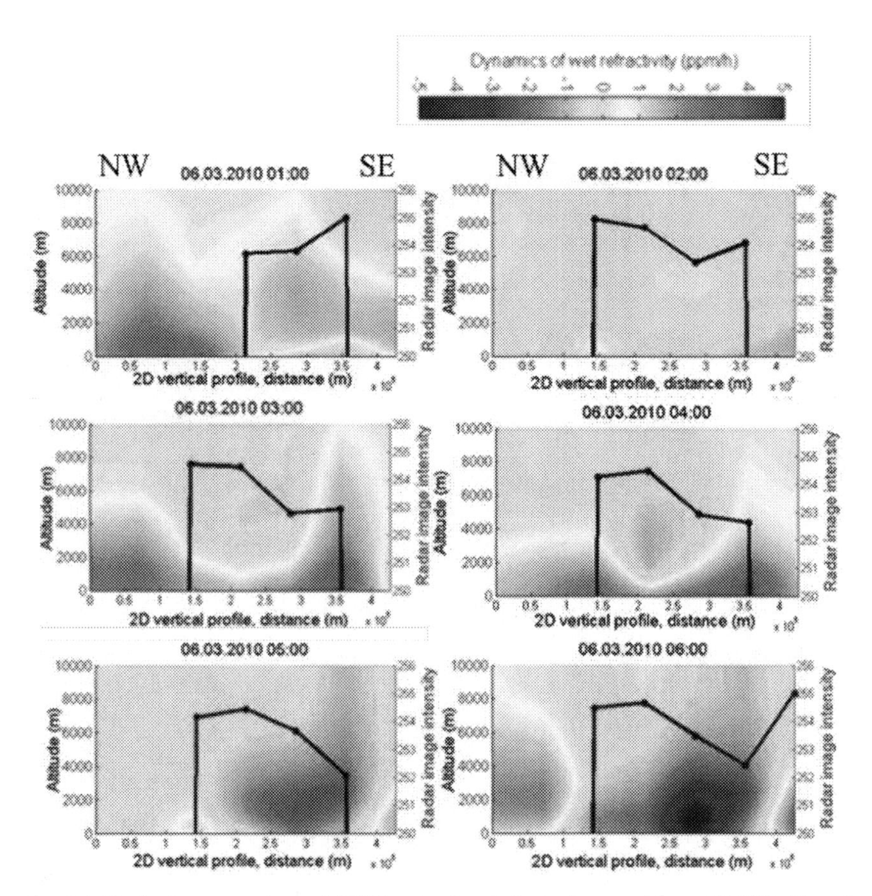

Fig. 6.9 Vertical cross section of GNSS wet refractivity (color map) from 22:00 UTC on 5 March to 06:00 on 6 March 2010. The *black line* is weather radar image intensity. *(From Zhang, K., Manning, T., Wu, S., Rohm, W., Silcock, D., & Choy, S. (2015). Capturing the signature of severe weather events in Australia using GPS measurements.* IEEE Journal of Selected Topics in Applied Earth Observations and Remote Sensing, *8(4), 1839–1847.)*

References

Bender, M., Stosius, R., Zus, F., Dick, G., Wickert, J., & Raabe, A. (2011). GNSS water vapour tomography – Expected improvements by combining GPS, GLONASS and Galileo observations. *Scientific Applications of Galileo and Other Global Navigation Satellite Systems - II, 47*(5), 886–897. https://doi.org/10.1016/j.asr.2010.09.011.

Brenot, H., Walpersdorf, A., Reverdy, M., Van Baelen, J., Ducrocq, V., Champollion, C., et al. (2014). A GPS network for tropospheric tomography in the framework of the Mediterranean hydrometeorological observatory Cévennes-Vivarais (southeastern France). *Atmospheric Measurement Techniques*, 7(2), 553–578.

Davis, J. L., Herring, T. A., Shapiro, I. I., Rogers, A. E. E., & Elgered, G. (1985). Geodesy by radio interferometry: Effects of atmospheric modeling errors on estimates of baseline length. *Radio Science, 20*(6), 1593–1607. https://doi.org/10.1029/RS020i006p01593.

Debye, P. J. W. (1929). *Polar molecules*. Dover Publications.

Elgered, G. (1993). Tropospheric radio-path delay from ground-based microwave radiometry. *Atmospheric Remote Sensing by Microwave Radiometry, 199*(3), 215–258.

Flores, A., Ruffini, G., & Rius, A. (2000). 4D tropospheric tomography using GPS slant wet delays. *Annales Geophysicae, 18*(2), 223–234. https://doi.org/10.1007/s00585-000-0223-7.

Gradinarsky, L. (2002). *Sensing atmospheric water vapor using radio waves: Studies of the 2, 3 and 4-D structure of the atmospheric water vapor using ground-based radio techniques comprising the global positioning system, microwave radiometry and very long baseline interferometry*.

Hirahara, K. (2000). Earth. *Planets and Space, 52*(11), 935–939. https://doi.org/10.1186/BF03352308.

Kruse, L. P. (2000). *Spatial and temporal distribution of atmospheric water vapor using space geodetic techniques*. Zurich: ETH.

Nolet, G. (1987). *Seismic wave propagation and seismic tomography* (pp. 1–23). Netherlands: Springer. https://doi.org/10.1007/978-94-009-3899-1_1.

Pollak, B. (1953). Experiences with planography* *from the Fort William sanatorium, Fort William, Ontario, Canada. *Diseases of the Chest, 24*(6), 663–669. https://doi.org/10.1378/chest.24.6.663.

Rüeger, J. M. (2002). *Refractive indices of light, infrared and radio waves in the atmosphere*. School of Surveying and Spatial Information Systems, University of New South.

Seko, H., Nakamura, H., Shoji, Y., & Iwabuchi, T. (2004). The meso-γ scale water vapor distribution associated with a thunderstorm calculated from a dense network of GPS receivers. *Journal of the Meteorological Society of Japan, 82*(1B), 569–586. https://doi.org/10.2151/jmsj.2004.569.

Troller, M., Bürki, B., Cocard, M., Geiger, A., & Kahle, H.-G. (2002). 3-D refractivity field from GPS double difference tomography. *Geophysical Research Letters, 29*(24), 2149. https://doi.org/10.1029/2002GL015982.

Ware, R., Alber, C., Rocken, C., & Solheim, F. (1997). Sensing integrated water vapor along GPS ray paths. *Geophysical Research Letters, 24*(4), 417–420. https://doi.org/10.1029/97GL00080.

Zhang, K., Manning, T., Wu, S., Rohm, W., Silcock, D., & Choy, S. (2015). Capturing the signature of severe weather events in Australia using GPS measurements. *IEEE Journal of Selected Topics in Applied Earth Observations and Remote Sensing, 8*(4), 1839–1847.

CHAPTER 7

Climate monitoring with GNSS (GNSS-C)

Climate and climate system components

IPCC AR5 (2014) defines the climate system as "the highly complex system consisting of five major components: the atmosphere, the hydrosphere, the cryosphere, the lithosphere, and the biosphere, and the interactions between them. The climate system evolves in time under the influence of its own internal dynamics and because of external forcings such as volcanic eruptions, solar variations, and anthropogenic forcings such as the changing composition of the atmosphere and land use change." The climate system components are often discussed in separation due to their different physical properties such as density, thermal capacity, thermal conductivity, reflectivity, mobility, etc., which also determines their different spatiotemporal behavior. The Earth's climate is also influenced by extraterrestrial factors such as solar radiation, Earth's orbit, and interplanetary interactions (Fig. 7.1).

Atmosphere

The Earth's atmosphere is a thin shell of air spreading to an altitude of 1000 km. Of particular importance for life on the Earth is the water cycle (hydrologic cycle) between the reservoirs and atmosphere, ocean, soils, groundwater, and ice (Fig. 7.2). On annual basis the evaporation from ocean and soils releases 423×10^{12} m^3 of water in the atmosphere. When the water vapor reaches its dew point, it condenses and releases energy that warms the atmosphere. Evaporation and condensation processes lead to rapid redistribution of energy into the atmosphere. The condensation results in cloud formation followed by precipitation. On an annual basis precipitation falling back to both ocean and soil is estimated to be 423×10^{12} m^3, i.e., it closes the cycle. The key role of water vapor in the Earth's energy balance has been summarized as "about 50% of the absorbed solar radiation on the Earth's surface is absorbed by the evaporation process, which results in cooling of the

Global Navigation Satellite System Monitoring of the Atmosphere
https://doi.org/10.1016/B978-0-12-819152-1.00008-8

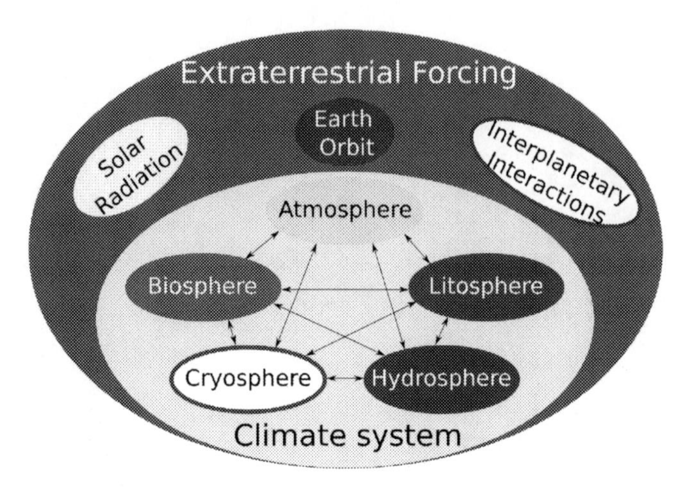

Fig. 7.1 Schematic representation of climate system components and external forcing. *(Courtesy: Tzvetan Simeonov.)*

surface" (Bengtsson, 2010). In addition to its dynamic role, water vapor is a very important greenhouse gas.

The greenhouse effect is of great importance to the climate of the planet (Paxi—The Greenhouse Effect, n.d.). When the shortwave radiation from the Sun reaches the Earth's surface, it is absorbed. The absorbed radiation causes heating of the surface and emission of long-wave infrared radiation back into the atmosphere. Greenhouse gases, such as water vapor, carbon dioxide, methane, etc., absorb long-wave infrared radiation of the Earth,

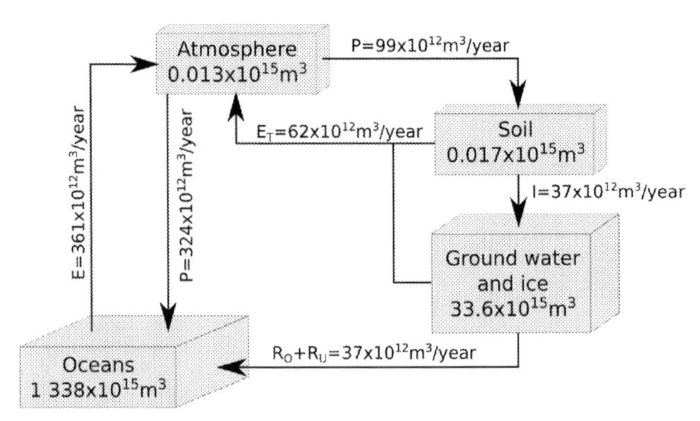

Fig. 7.2 The hydrological cycle of the Earth system. *(From Derivation and analysis of atmospheric water vapor and soil moisture from ground-based GNSS stations. (2021). TU Berlin.)*

thus keeping the Earth's surface warm. Without greenhouse gasses, the Earth's atmosphere would be on average between 30°C and 35°C colder. Water vapor contributes most to the greenhouse effect—about 75% of the Earth's greenhouse effect (Kondratyev, 1972) is due to water vapor present in the troposphere. According to (Buehler et al., 2006) an increase of 20% of water vapor in tropical regions will result in approximately the same effect as a doubling of carbon dioxide concentration. The amount of water vapor depends on the air temperature. As a result of global warming the moisture-holding capacity of the air increases, which in turn increases the amount of water vapor. Assuming constant relative humidity, the vertically integrated water vapor increase will be approximately 7% for every 1 K temperature increase (Trenberth, Dai, Rasmussen, & Parsons, 2003). Mears, Santer, Wentz, Taylor, and Wehner (2007) found an increase in water vapor by 5%–7% for each 1 K. (Ross & Elliott, 1996, 2001) analyzed radiosonde data for the period 1973–1995 and identified positive water vapor trends over China, the Pacific Islands, and the United States. The main disadvantage of using radiosondes for trend estimation is systematic errors caused by calibration and/or replacement of measuring instruments leading to inhomogeneous time series (Wang & Zhang, 2008).

Hydrosphere

The hydrosphere is the largest component of the climate system, which consists of all water bodies such as oceans, seas, lakes, ponds, rivers, and streams. The global ocean is the interconnected body of saline water that encompasses polar to equatorial climate zones and covers 71% of the Earth's surface. The ocean contains about 97% of the Earth's water, supplies 99% of the Earth's biologically habitable space, and provides roughly half of the primary production on Earth (Gattuso, Abram, & Hock, 2019). The oceans absorb and transport heat globally and are a fundamental climate regulator on seasonal to millennial time scales. Seawater has a heat capacity four times larger than air and holds large quantities of dissolved carbon. Heat, water, and biogeochemically relevant gases [oxygen (O_2) and carbon dioxide (CO_2)] exchange at the air-sea interface. Ocean currents and mixing caused by winds, tides, wave dynamics, density differences, and turbulence redistribute these gases throughout the global ocean. Fig. 7.3 shows the climate change-related effects in the ocean namely (1) sea level rise, (2) increasing ocean heat content and marine heat waves, (3) ocean deoxygenation, and (4) ocean acidification (Gattuso et al., 2019).

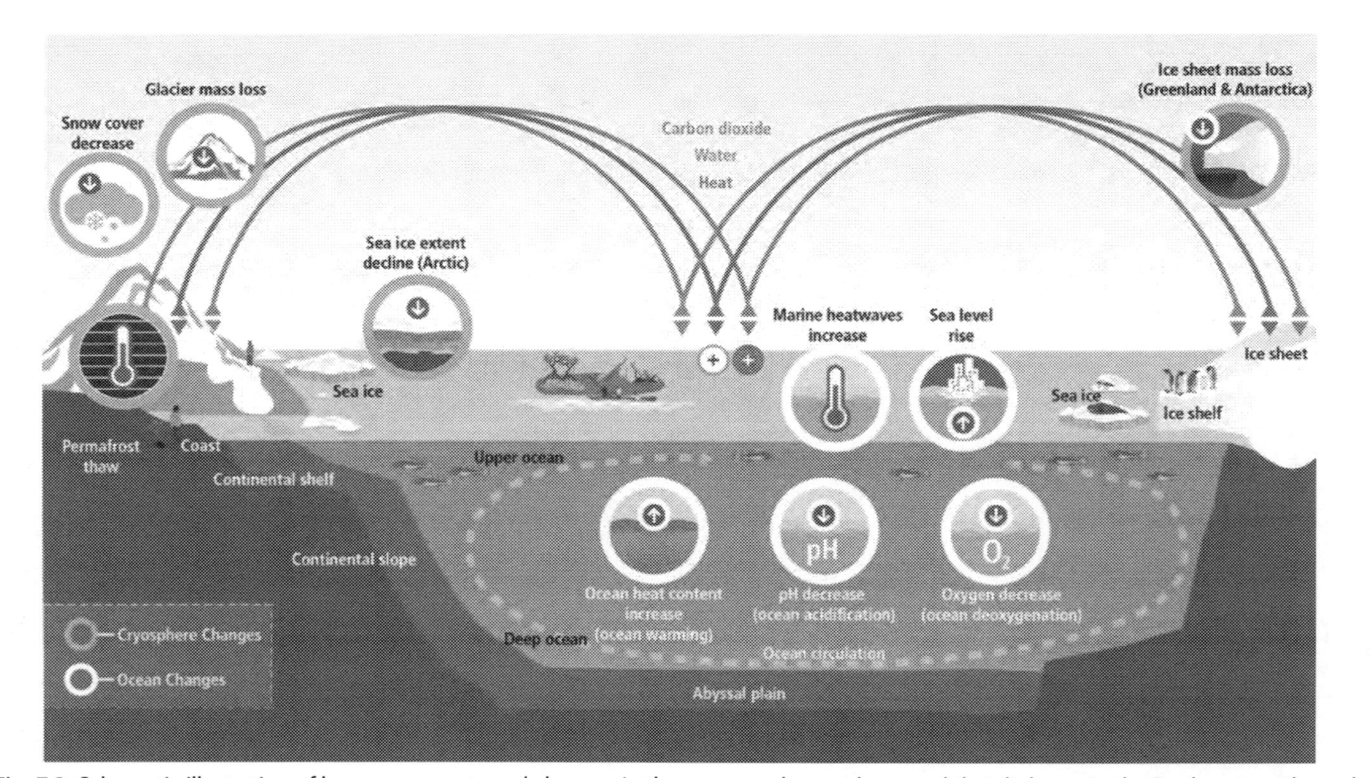

Fig. 7.3 Schematic illustration of key components and changes in the ocean and cryosphere, and their linkages in the Earth system through the movement of heat, water, and carbon. *(Courtesy: IPCC, 2019: Summary for Policymakers. (2019). In IPCC Special Report on the Ocean and Cryosphere in a Changing Climate. https://report.ipcc.ch/srocc/pdf/SROCC_FinalDraft_FullReport.pdf.)*

Cryosphere

The cryosphere refers to frozen components of the climate system such as snow, glaciers, ice sheets, ice shelves, icebergs, sea ice, lake ice, river ice, permafrost, and seasonally frozen ground (Gattuso et al., 2019). Snow is common in polar and mountain regions and it melts seasonally or transforms into ice layers that build glaciers and ice sheets. Snow feeds groundwater and river runoff (Fig. 7.3) and together with glacier melt, causes natural hazards like avalanches and rain-on-snow flood events (Gattuso et al., 2019). Snow affects the Earth's energy budget by reflecting solar radiation (albedo effect, see "Lithosphere and biosphere" section), and influences the temperature of underlying permafrost. Ice sheets and glaciers are land-based ice, built up by accumulating snowfall on their surface. Currently, around 10% of Earth's land area is covered by glaciers or ice sheets with Greenland and Antarctic ice sheets (a) being the largest ice bodies. Ice sheets flow outward from their dome-like centers (b) and push ice outward until they encounter the ocean, or where the climate is warm enough to melt the ice faster than the combined flow rate and winter snowfall (National Snow and Ice Data Center (NSIDC), State of the Cryosphere, 2019). For much of Greenland and Antarctica, ice flow terminates at the ocean, as a tidewater glacier or an ice tongue or ice shelf. Ice sheets and glaciers that lose more ice than they accumulate contribute to global sea level rise. Ice shelves are extensions of ice sheets and glaciers that float in the surrounding ocean (Gattuso et al., 2019). Sea ice forms from the freezing of seawater, and sea ice floating on the ocean surface is further thickened by snow accumulation (Gattuso et al., 2019). Sea ice has critical functions in the climate system such as (1) regulating climate by reflecting solar radiation; (2) inhibiting ocean-atmosphere exchange of heat, momentum, and gases; and (3) supporting global deep ocean circulation via dense cold and salty water formation (Gattuso et al., 2019). Permafrost is soil or rock containing ice and frozen organic material that remains at or below 0°C for at least two consecutive years (Gattuso et al., 2019). It occurs on land in polar and high-mountain areas with thickness ranging from 1 to 1000 m. It usually occurs beneath an active layer, which thaws and freezes annually. Permafrost thaw can cause hazards, including ground subsidence or landslides, and influence global climate through the emission of greenhouse gases from microbial breakdown of previously frozen organic carbon. Changes in the cryosphere include (1) the decline of Arctic sea ice extent, (2) Antarctic and Greenland ice sheet mass loss (Fig. 7.4), (3) glacier mass loss, (4) permafrost thaw, and (5) decreasing snow cover extent (Gattuso et al., 2019).

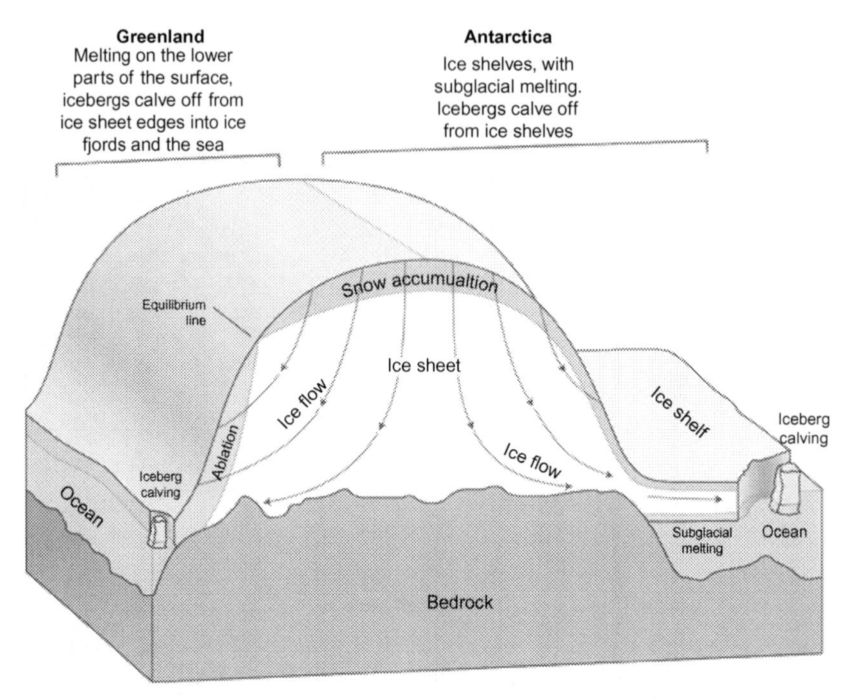

Fig. 7.4 A cross-section of an ice sheet. *(From National Snow and Ice Data Center: https://nsidc.org/cryosphere/sotc/ice_sheets.html.)*

Lithosphere and biosphere

The lithosphere is the upper layer of the solid Earth, both continental and oceanic, which comprises all crustal rocks and the cold, mainly elastic part of the uppermost mantle. Volcanic and hydrothermal vents in the lithosphere emit volcanic ash and CO_2 and reduce warming by a process called global dimming. Large volcanic eruptions affect climate on short time scales of a few years to decades. The International Panel for Climate Change (IPCC) defines biosphere as the part of the Earth system comprising all the living organisms in the atmosphere, on land (terrestrial biosphere), and in the ocean (marine biosphere), including derived dead organic matter (litter, soil organic matter, and oceanic detritus) (Solomon, Manning, Marquis, & Qin, 2007). One of the most important roles of terrestrial vegetation for the Earth's climate is its capacity to reflect the incoming sunlight back into the atmosphere (albedo). Vegetation usually has a lower albedo than soil, in particular, lower than that of deserts (Fig. 7.5). Albedo maximum is also observed at high latitudes because of the presence of snow and ice. Greenhouse gases exchange between the land and the atmosphere is driven

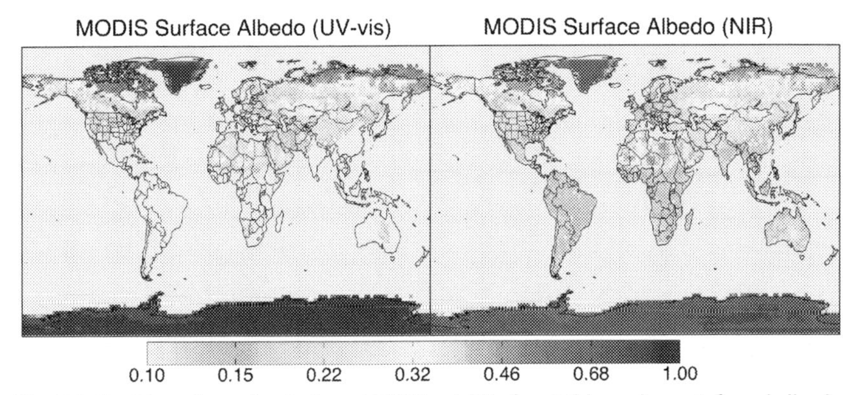

Fig. 7.5 Earth's surface albedo from MODIS satellite for visible and near infrared albedo. *(From Heald, C. L., Ridley, D. A., Kroll, J. H., Barrett, S. R. H., Cady-Pereira, K. E., Alvarado, M. J., & Holmes, C. D. (2014). Contrasting the direct radiative effect and direct radiative forcing of aerosols. Atmospheric Chemistry and Physics, 14, 5513–5527. https://doi.org/10.5194/acp-14-5513-2014.)*

mainly by the balance between photosynthesis and plant respiration, and by the decomposition of soil organic matter by microbes (IPCC, 2019). The conversion of atmospheric CO_2 into organic compounds by plant photosynthesis, known as terrestrial net primary productivity, is the source of plant growth, food for living organisms, and soil organic carbon. Due to strong seasonal patterns of growth, terrestrial ecosystems in the Northern Hemisphere are largely responsible for the seasonal variations in global atmospheric CO_2 concentrations. In addition to CO_2, soils emit methane (CH4) and nitrous oxide (N_2O). Soil temperature and moisture strongly affect microbial activities and resulting greenhouse gas fluxes. The biosphere climate change-related effects include changes in albedo, strength of land carbon sinks or sources, soil moisture, and plant phenology.

Climate variability, climate change indices, and teleconnections

Global atmospheric circulation is characterized by several "preferred" modes of variability. (Trenberth, Smith, Qian, Dai, & Fasullo, 2007) studied the global atmospheric circulation and the emergence of teleconnection patterns. The regional climate at different points of the Earth can vary in counter phase as a result of teleconnection patterns that modulate the location and depth of cyclones and flows of heat, moisture, and angular momentum to the pole. Each teleconnection pattern consists of a specific spatial configuration of the geophysical field anomalies and a time series index characterizing its amplitude and phase. In the Northern Hemisphere, mean atmospheric circulation is most active in winter, thus it is considered representative of interannual changes. Because the temperature variability is greater outside the tropics, these regions dominate the hemispheric temperature anomaly estimates (Barnett, 1978; Hurrell, 1995; Wallace, Zhang, & Renwick, 1995). Teleconnection patterns are also best expressed in winter and beyond the tropical latitudes. Winter extratropical teleconnection patterns are closely related to the temperature anomaly in the Northern Hemisphere.

North Atlantic oscillation

Research shows that North Atlantic Oscillation (NAO) is related to variability in temperatures, rainfall, and the stormy weather frequency in Europe in winter. The NAO links the relatively slow variability and inertia of Atlantic waters with atmospheric circulation. The forecast of atmospheric circulation is made by comparing the temperature of surface waters in the North Atlantic in May with the average atmospheric conditions from December to February of the following winter season. Thus, based on the observed surface water temperature of the North Atlantic, the NAO index forecast the coming winter season. One of the NAO definitions is the difference in atmospheric sea level pressure, in winter, between the Azores and Iceland. By modulating the intensity of the Icelandic minimum and the Azores maximum, the NAO modulates the intensity of the westerly flow over the North Atlantic and Europe. The effect is strongest in the winter months (December to March). The positive NAO phase (Fig. 7.6A) leads to warming in much of Eurasia and Eastern North America, and cooling in Greenland, the Eastern Mediterranean, the Middle East, North Africa, Central America, Alaska, and Western North America. Fig. 7.6A shows the atmospheric circulation in the North Atlantic during the positive NAO phase. The correlation of NAO and temperatures and precipitation is shown in Fig. 7.6B and C. The NAO explains about 31% of the annual variability in surface temperature of the Northern Hemisphere's in winter (Hurrell, 1995).

El Niño and Southern Oscillation Index (ENSO)

El Niño is the result of the interaction between the atmosphere and the ocean and its phases are related to the change in Pacific Ocean surface temperature (SST—sea surface temperature) in the tropical region between the International Date Line and the South American coast (Fig. 7.7A). The SST

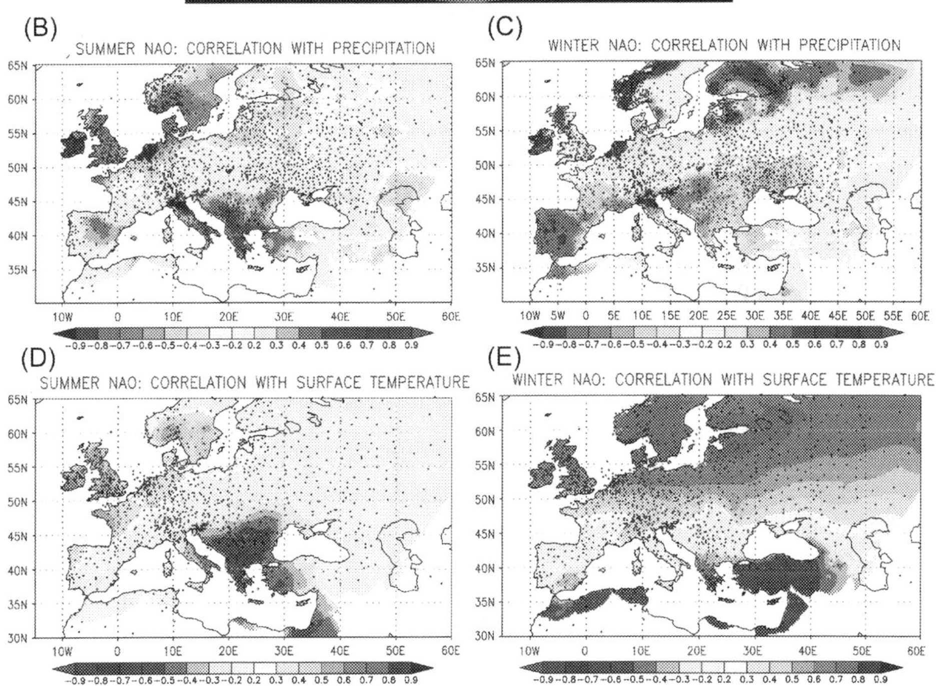

Fig. 7.6 (A) Illustration of a positive NAO index. (B) Global correlation coefficient between the NAO and temperature *(upper panel)* and precipitation *(lower panel)* for December–March and (C) for Europe only. *(From (A) Positive NAO Index, https://www.ldeo.columbia.edu/res/pi/NAO/, (B) https://www.ldeo.columbia.edu/res/pi/NAO/intro/correlations.html. Bladé, I., Liebmann, B., Fortuny, D., & van Oldenborgh, G. J. (2012) observed and simulated impacts of the summer NAO in Europe: implications for projected drying in the Mediterranean region.* Climate Dynamics, *39(3–4), 709–727. https://citeseerx.ist.psu.edu/viewdoc/download?doi=10.1.1.368.3679&rep=rep1&type=pdf.)*

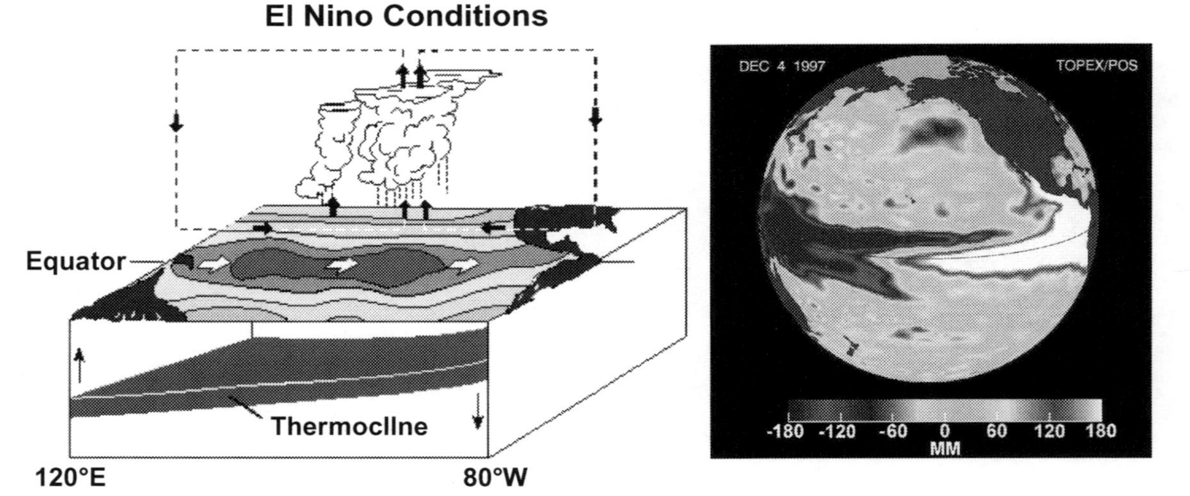

Fig. 7.7 (A) Schematic representation of ocean circulation and atmosphere at El Nino. (B) Positive anomaly *(in white)* of seawater temperature in the Pacific Ocean off the coast of South America during El Nino in 1997. *((A) NOAA, What is El Niño? Available from: https://www.pmel.noaa.gov/elnino/what-is-el-nino, (B) El Niño: 1997–1998 vs. 2015–2016. Available from: https://sealevel.jpl.nasa.gov/data/el-nino-la-nina-watch-and-pdo/el-nino-2015/.)*

defines the exchange of heat between the atmosphere and the ocean, hence their circulation, and ultimately affects the average global temperature. The positive phase, El Niño, leads to warming in the west coast of South America, Southeast Africa, Southeast Asia, India, Northwestern North America, Japan, Central America, the Caribbean, and Central South America, and cooling in the Central South and Western Pacific. Fig. 7.7B shows the temperature anomaly on the surface of the Pacific Ocean in 1997, characterized by the strong El Niño. The second variability pattern in the Southern Hemisphere is the Southern Oscillation Index (SOI). SOI is defined as the standardized difference between the standardized monthly atmospheric pressure values in the cities of Darwin (Australia) and Tahiti (French Polynesia). SOI varies between years and decades (Wang & Zhang, 2008) and explains about 16% of the annual variance of surface temperature in the Northern Hemisphere in winter (Hurrell, 1995). The positive SOI phase corresponds to the positive temperature anomaly in the Northern Hemisphere and the negative phase to negative. The absolute value of the temperature anomaly in the Northern Hemisphere, as a result of the SOI phases, is estimated to be about 0.1°C (Jones, 1989). El Niño and the Southern Oscillation are known as ENSO.

Climate system reanalysis

An accurate description of the present state of the climate system is particularly important for its forecasting. For this purpose, a combination of observations and numerical models are used to make the most accurate description of climate in the last century—climate reanalysis. Leading meteorological centers in Europe, the United States, and Japan are creating climate reanalysis that include the atmosphere, land, ocean, cryosphere, and carbon cycle. The work of creating a reanalysis includes, as a first step, the collection, preparation, and evaluation of climate observations, from the earliest direct observations made by meteorological observers to modern, high-resolution satellite observations. The second step is to assimilate the observations in numerical weather models to ensure the best possible time consistency of the reanalysis. Because reanalyses are made with the best observation assimilation systems developed for numerical weather models, they are suitable for studies of long-term climate variability. Reanalysis products are increasingly used in many areas that require monitoring of the state of the atmosphere and/or terrestrial and oceanic surfaces.

Global climate system reanalysis

This section presents the global realities of one of the leading centers—the European Center for Medium-Term Weather Forecast (ECMWF Reanalysis, n.d.). ERA-15 is the first global reanalysis of the ECMWF and covers the period 1979–1993. Its realization began in the early 1980s and ended in 1995. The second reanalysis ERA-40 extends the period to 40 years 1957–2002. The third-generation ERA-Interim reanalysis (1979–2018) includes the period after 1989 with a large number of satellite observations. The fourth-generation ERA-20C relaunch, covering the period 1900–2010, is the first reanalysis of the 20th-century Atmospheric-Ocean-Land Climate System (ERA-20C, n.d.). The ERA-20C's products cover the spatiotemporal dynamics of the atmosphere with 91 vertical levels from the Earth's surface to 0.01 hPa, four soil layers, and ocean waves with 25 frequencies and 12 directions. The horizontal resolution is 125 km with 3 hourly products. The method of assimilation of observations is 4D-Var with variational correction for ground pressure observations. Analyses provide initial conditions for subsequent forecasts that serve as a basis for subsequent analysis. Table 7.1 presents the ERA-20C implementation scheme with forecast initiated at 06 UTC and forecast steps.

The fifth-generation ECMWF reanalysis ERA5 is in production and covers 1950–present (ERA5, n.d.). ERA5 has a horizontal resolution of 31 km and 137 vertical levels from the Earth's surface up to 80 km (Table 7.2). ERA-15, ERA-40, and ERA-Interim products are widely used by the member states and end users. They are also an important part of ECMWF's core activities in particular they are used to evaluate long-term numerical simulations as well as to support the development of seasonal forecasts and are at the heart of the widely used extreme weather index. ERA5 is contributing to the Copernicus climate change service (Copernicus Program, n.d.). A review of the global climate reanalysis is presented by Fujiwara et al. (2017). Table 7.2 is a summary of the horizontal and vertical resolution of the ECMWF—ERA reanalysis, Japan Meteorology Agency—JRA reanalysis, NASA—MERRA reanalysis, and NCEP—R1 to 20CR reanalysis. The state-of-the-art reanalysis assimilated a large number of observations including conventional (Fig. 7.8) and satellite observations (Fig. 7.9).

Regional reanalysis—German Weather Service (DWD)

The leading European Meteorological Services are currently producing regional renaissances and one of them is the Regional Reality System

Table 7.1 ERA-20C implementation scheme.

Step	3	6	9	12	15	18	21	24	27
Valid time, for instantaneous forecast parameters	09UTC	12UTC	15UTC	18UTC	21UTC	00UTC next day	03UTC next day	06UTC next day	09UTC next day
Accumulation period, for accumulated forecast parameters	06UTC to 09UTC	06UTC to 12UTC	06UTC to 15UTC	06UTC to 18UTC	06UTC to 21UTC	06UTC to 00UTC next day	06UTC to 03UTC next day	06UTC to 06UTC next day	06UTC to 09UTC next day

The forecasts are daily initialized at 06 UTC and are +3, +6, +9, +12, +15, +18, +21, +24 and +27 h (row 1).

Source: https://www.ecmwf.int/en/forecasts/datasets/reanalysis-datasets/era-20c.

Table 7.2 Global climate reanalysis table.

Reanalysis name	Model (year)	Horizontal grid resolution	Vertical levels
ERA-40	IFS Cycle 23r4 (2001)	~125 km	60
ERA-Interim	IFS Cycle 31r2 (2007)	~79 km	60
ERA-20C	IFS Cycle 38r1 (2012)	~125 km	91
ERA5	IFS Cycle 41r2 (2016)	~30 km	137
JRA-25	JMA GSM (2004)	1.125°	40
JRA-55	JMA GSM (2009)	~55 km	60
MERRA	GEOS 5.0.2 (2008)	0.5° latitude × 0.66° longitude	72
MERRA-2	GEOS 5.12.4 (2015)	0.5° latitude × 0.625° longitude	72
R1	NCEP MRF (1995)	1.875°	28
R2	Modified MRF (1998)	1.875°	28
CFSR	NCEP CFS (2007)	0.3125°	64
CFSv2	NCEP CFS (2011)	0.2045°	64
20CR	NCEP GFS (2008)	1.875°	28

Column: (1) name of reanalysis, (2) name of numerical model version and year, (3) horizontal resolution, (4) number of vertical levels.
Fujiwara, M., Wright, J. S., Manney, G. L., Gray, L. J., Anstey, J., Birner, T., Davis, S., Gerber, E. P., Harvey, V. L., & Hegglin, M. I. (2017). Introduction to the SPARC Reanalysis Intercomparison Project (S-RIP) and overview of the reanalysis systems. *Atmospheric Chemistry and Physics*, *17*(2), 1417–1452.

developed by the German Meteorological Service based on the COSMO numerical weather forecast model (COSMO-REA*6*, 2021). COSMO-REA6 covers continental Europe and has a horizontal resolution of 6 km, 40 vertical levels, and boundary conditions from ERA-Interim. Observations are assimilated by the nudging method and include standard observations from aerological sounding (PILOT and TEMP), regular observations from passenger aircraft instruments (AIREP, AMDAR, ACARS), wind profiles (wind profiler), and surface observations (SYNOP, SHIP, DRIBU). In addition, snow cover, sea surface temperature, and soil moisture are assimilated. Fig. 7.10 shows the sequence of reanalysis procedures. The initial reanalysis period covers 1995–2017 and is projected to expand over time.

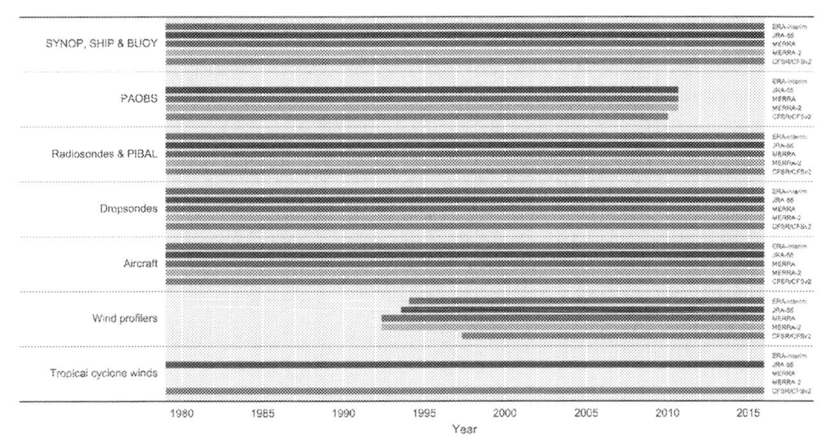

Fig. 7.8 Types of direct observations assimilated to ERA-Interim *(blue; dark gray in print version)*, JRA-55 *(purple; light gray in print version)*, MERRA and MERRA-2 *(dark and light red; light gray in print version)* and CFSR *(green; dark gray in print version)*. *(From Fujiwara, M., Wright, J. S., Manney, G. L., Gray, L. J., Anstey, J., Birner, T., Davis, S., Gerber, E. P., Harvey, V. L., & Hegglin, M. I. (2017). Introduction to the SPARC Reanalysis Intercomparison Project (S-RIP) and overview of the reanalysis systems.* Atmospheric Chemistry and Physics, 17*(2), 1417–1452.)*

Monitoring long-term changes in atmospheric water vapor with GNSS

GNSS reprocessing campaign of tropospheric products

An important element of climate monitoring is to use consistently processed long time series. In order to use GNSS for monitoring long-term water vapor variability as a first step consistent processing of the GNSS stations is required. Fig. 7.11 presents the state-of-the-art GNSS processing sequence (GNSS-reprocessing or GNSS-repro). To produce GNSS-repro it is necessary to use the original GNSS observations (RINEX files), as well as information about changes of the equipment or moving the station (site logs). The stations, for which this information is available, are processed by calculating coordinates and tropospheric products. For this purpose, predefined parameters for height above the horizon (elevation cutoff angle), a priori models for the troposphere (mapping function), frequency of observations, etc., are used. After calculating tropospheric products such as zenith total delay (ZTD), it is necessary to perform a screening procedure and calculate the integrated water vapor (IWV). The resulting IWV time series undergo a homogenization procedure to detect discontinuities and correct systematic errors.

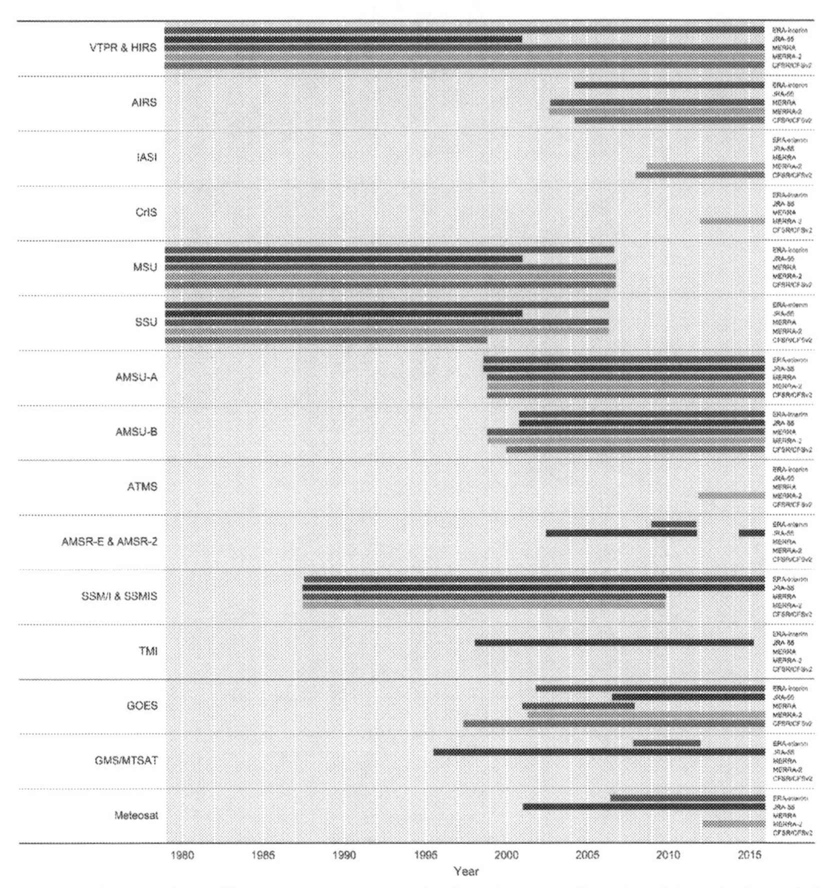

Fig. 7.9 Types of satellite observations assimilated in ERA-Interim *(blue; dark gray in print version)*, JRA-55 *(purple; light gray in print version)*, MERRA and MERRA-2 *(dark and light red; light gray in print version)* and CFSR *(green; dark gray in print version)*. *(From Fujiwara, M., Wright, J. S., Manney, G. L., Gray, L. J., Anstey, J., Birner, T., Davis, S., Gerber, E. P., Harvey, V. L., & Hegglin, M. I. (2017). Introduction to the SPARC Reanalysis Intercomparison Project (S-RIP) and overview of the reanalysis systems.* Atmospheric Chemistry and Physics, 17*(2), 1417–1452.)*

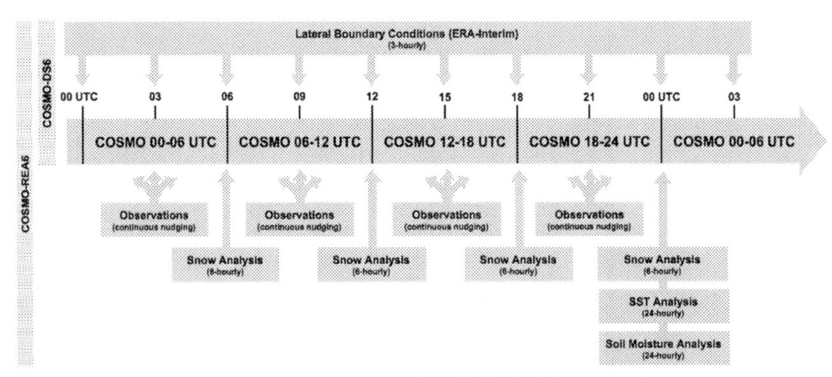

Fig. 7.10 Schematic representation of COSMO-REA6 implementation steps. *(COSMO-REA6. (2021). http://reanalysis.meteo.uni-bonn.de/?COSMO-REA6.)*

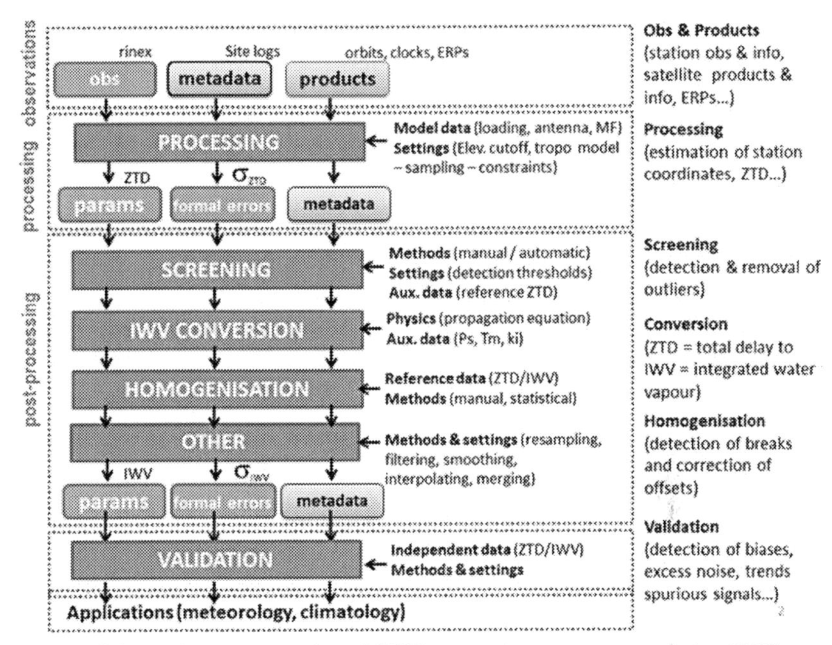

Fig. 7.11 Schematic representation of GNSS processing sequence to derive GNSS-repro time series of ZTD and IWV. *(From Jones, J., Guerova, G., Douša, J., Dick, G., de Haan, S., Pottiaux, E., & van Malderen, R. (2019). Advanced GNSS tropospheric products for monitoring severe weather events and climate. COST Action ES1206 Final Action Dissemination Report, 563.)*

List of global GNSS-repro of tropospheric products: (IGS-Repro1, n.d.; TIGA Reprocessing Campaign, n.d.; GRUAN Reprocessing Campaign, n.d.). European reprocessing is conducted within EUREF: (EPN-Repro1, n.d.; EPN-Repro2, n.d.). The GNSS-repro water vapor time series are used to evaluate (1) global and regional atmospheric reanalysis and climate models, (2) diurnal cycle of water vapor in atmospheric reanalysis and climate models, and (3) variability and linear trends in different climatic zones.

Global GNSS research

The first global intercomparison between GNSS reprocessed products and NCEP-R2 reanalysis (see Table 7.2) was conducted by (Vey et al., 2010). High correlation between seasonal and annual variations with a correlation coefficient of 0.9 or higher in midlatitudes is reported (Fig. 7.12). In the tropics and Antarctica, the correlation coefficient is in the range 0.6–0.9, with a tendency for R2 reanalysis to decrease in water vapor by 40% and 25%,

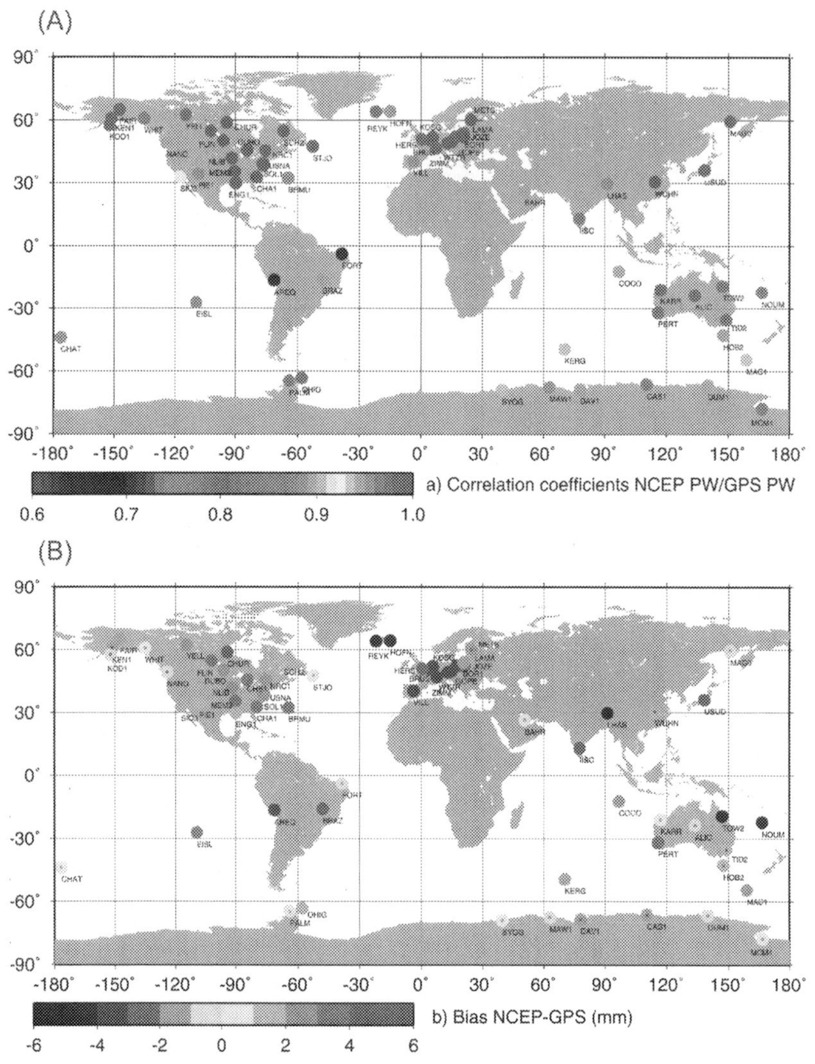

Fig. 7.12 Comparison of GNSS water vapor and NCEP-R2 reanalysis: (A) correlation coefficient and (B) mean difference. *(From Vey, S., Dietrich, R., Rülke, A., Fritsche, M., Steigenberger, P., & Rothacher, M. (2010). Validation of precipitable water vapor within the NCEP/DOE reanalysis using global GPS observations from one decade.* Journal of Climate, *23(7), 1675–1695.)*

respectively. A recent study, in the framework of the COST Action GNSS4 SWEC, compares global time series of water vapor from IGS repro1 with ERA-Interim and MERRA-2 reanalysis for the period 1995–2010 (Parracho, Bock, & Bastin, 2018). As shown in Fig. 7.13 ERA-Interim is more humid in the extratropics and drier in the tropics (0.5–1 kg/m^2).

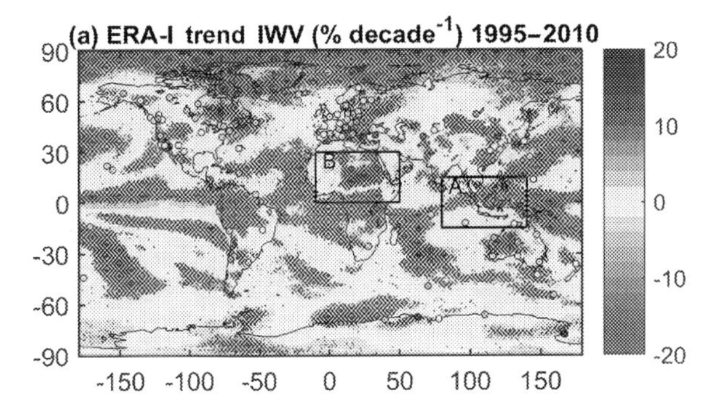

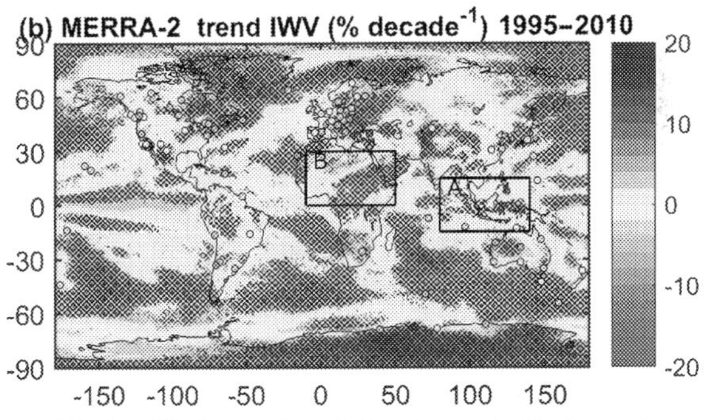

Fig. 7.13 Mean IWV for (A) December–February and (B) June–August from GNSS *(circles)* and ERA-Interim (color map). Relative IWV variability for (C) December–February and (D) June–August. *(From Parracho, A. C., Bock, O., & Bastin, S. (2018). Global IWV trends and variability in atmospheric reanalysis and GPS observations.* Atmospheric Chemistry and Physics, 18*(22), 16,213–16,237. https://acp.copernicus.org/articles/18/16213/2018/.)*

Compared to MERRA-2, ERA-Interim has lower water vapor values in the oceans and tropical areas. The linear trends of IWV and surface temperature in ERA-Interim and MERRA-2 are in good agreement (Fig. 7.14). Areas with positive trends in IWV usually correlate with the increase in surface temperature. MERRA-2 tends to overestimate the humidity globally when compared to ERA-Interim. The linear trends between the two reanalyses differ for Antarctica and most of the Southern Hemisphere, as well as Central and North Africa. There are few terrestrial observations in these areas and differences are expected. The interannual and decade-long variations in water vapor are found to be associated with changes in atmospheric circulation, especially in the desert regions of North Africa and Western Australia, which contribute to uncertainty in estimated trends, especially

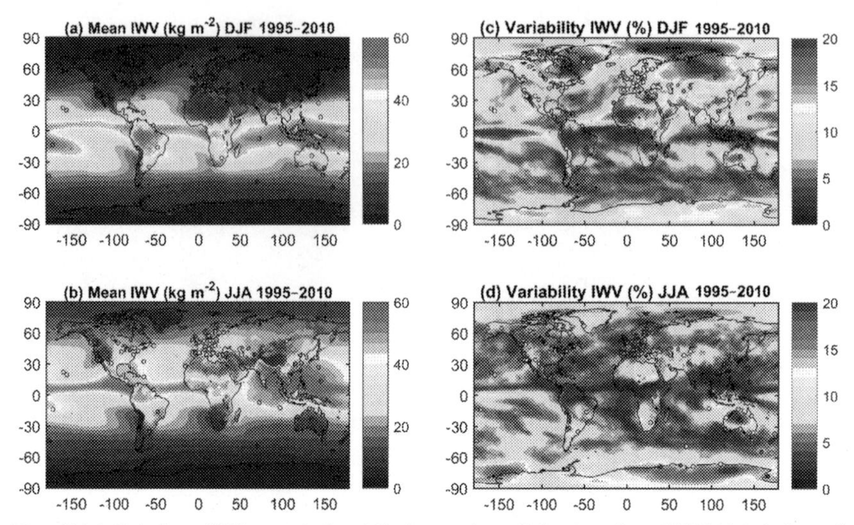

Fig. 7.14 Relative IWV trend for (A) December–February for GNSS *(circles)* and ERA-Interim (color map) and (B) the same as (A) but for MERRA-2. *(From Parracho, A. C., Bock, O., & Bastin, S. (2018). Global IWV trends and variability in atmospheric reanalysis and GPS observations.* Atmospheric Chemistry and Physics, *18(22), 16213–16237. https://acp.copernicus.org/articles/18/16213/2018/.)*

over a short period of 15 years. For these regions, the Clausius-Clapeyron equation does not agree with annual and decade fluctuations.

Regional GNSS research

Regional Climate Models (RCMs) with a horizontal resolution of 10–50 km are used as complementary data sources to global models. Numerical simulations with RCMs reproduce many features of the regional climate. A comparison of eight RCMs, with the same initial conditions for the western part of the Arctic Ocean, showed that they simulate well the monthly and daily amplitudes of the IWV, but with significant differences between the individual models. The first regional intercomparison between GNSS IWV and RCM is conducted with the regional Rossby Center Atmosphere (RCA) climate model developed by the Swedish Meteorological and Hydrological Institute. ERA-Interim is used to initialize RCA's model. The numerical simulations with RCA over a 14-year period are compared with GNSS observations from 99 stations in Europe. As shown in Fig. 7.15 the mean IWV difference varies in the range $-0.50\,kg/m^2$ to $+1.09\,kg/m^2$ for RCA-GNSS and $-0.21\,kg/m^2$ to $+1.12\,kg/m^2$ for

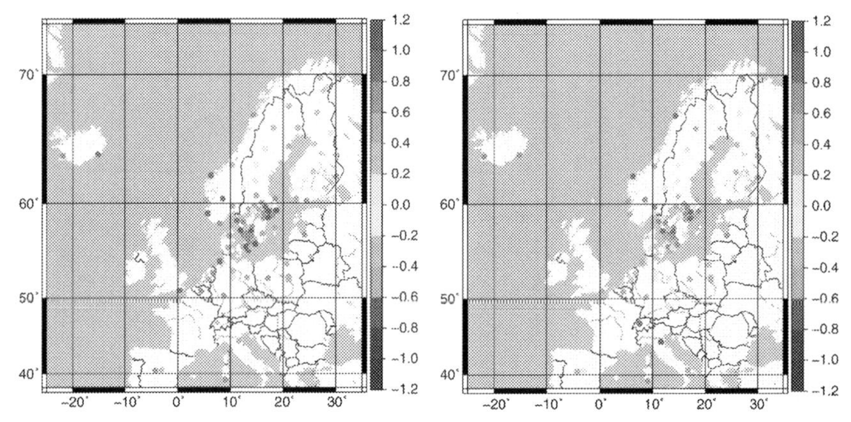

Fig. 7.15 Mean IWV difference: (A) RCA minus GNSS and (B) ERA-Interim minus GNSS. *(From Ning, T., Elgered, G., Willén, U., & Johansson, J. M. (2013). Evaluation of the atmospheric water vapor content in a regional climate model using ground-based GPS measurements.* Journal of Geophysical Research: Atmospheres, 118*(2), 329–339. https://agupubs.onlinelibrary.wiley.com/doi/full/10.1029/2012JD018053.)*

ERA-Interim-GNSS. Note the similar patterns in the mean IWV difference for both comparisons. A comparison of diurnal IWV cycle for the summer months (Fig. 7.16) shows very good agreement between GNSS and RCA during the day and wet bias in RCA over the night. The IWV amplitude in ERA-Interim is small and the mean value is high both for the nighttime and the daytime.

In Fig. 7.17 the peak time of the diurnal IWV cycle for the summer months is shown. Visual comparison between Fig. 7.17A and B shows a high correlation between GNSS and RCA. The RCA captures the geographical variations from west to east, with later peaks in the afternoon further east, and the late night and early morning peaks along the east coast of Sweden (Ning, Elgered, Willén, & Johansson, 2013). The IWV histogram for GNSS gives a peak between 16 and 19 LST (Fig. 7.17C) while the RCA peak is at 18 LST (Fig. 7.17D).

Linear trends of GNSS ZTD

Within the European Permanent Network second reprocessing campaign (EPN-repro2), five analysis centers have homogeneously processed the EPN network for the period 1996–2014. The EPN-repro2 includes coordinates and other metadata, individual and combined tropospheric products (Pacione, Araszkiewicz, Brockmann, & Dousa, 2017). EPN-repro2 data

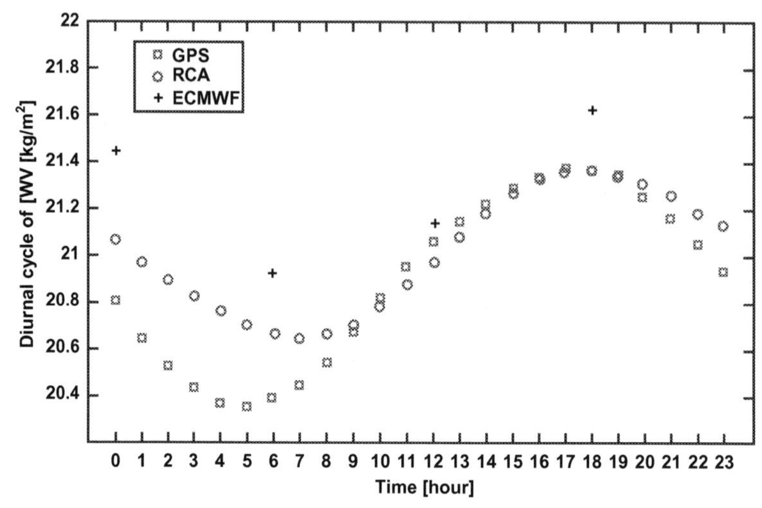

Fig. 7.16 Diurnal IWV cycle for GNSS *(red squares; dark gray print version)*, RCA *(blue dots; dark gray in print version)* and ERA-Interim *(black crosses). (From Ning, T., Elgered, G., Willén, U., & Johansson, J. M. (2013). Evaluation of the atmospheric water vapor content in a regional climate model using ground-based GPS measurements.* Journal of Geophysical Research: Atmospheres, 118*(2), 329–339. https://agupubs.onlinelibrary.wiley.com/doi/full/10.1029/2012JD018053.)*

recording can be used as a starting point for a variety of scientific applications. It provides great potential for monitoring atmospheric vapor trends and variability as well as intercomparison with regional climate models.

For the five EPN stations listed in Table 7.3 with long time series, ZTD trends are calculated using (1) EPN-repro2, (2) EPN-repro1, (3) IGS-repro1, (4) radiosonde, and (5) ERA-Interim data. All datasets are filtered for outliers using three sigma criteria and ZTD breakpoints, related to antenna change are homogenized. No homogenization is performed for the radiosonde. The least squares estimation method is used to evaluate trends and the seasonal components. The ZTD trends (Fig. 7.18) for the three GNSS ZTD datasets are consistent with the root mean square of 0.02 mm/year. The best agreement is found for ONSA (RMS = 0.04 mm/year) and WTZR (RMS = 0.02 mm/year). For PENC there is good agreement with ERA-Interim (0.05 mm/year), but a large discrepancy with the radiosonde (−0.31 mm/year). This large discrepancy is likely due to the distance between GNSS and radiosonde station (40.7 km) and the lack of homogenization. For the five stations better agreement is found with respect to the ERA-Interim (RMS = 0.11 mm/year) than with respect to

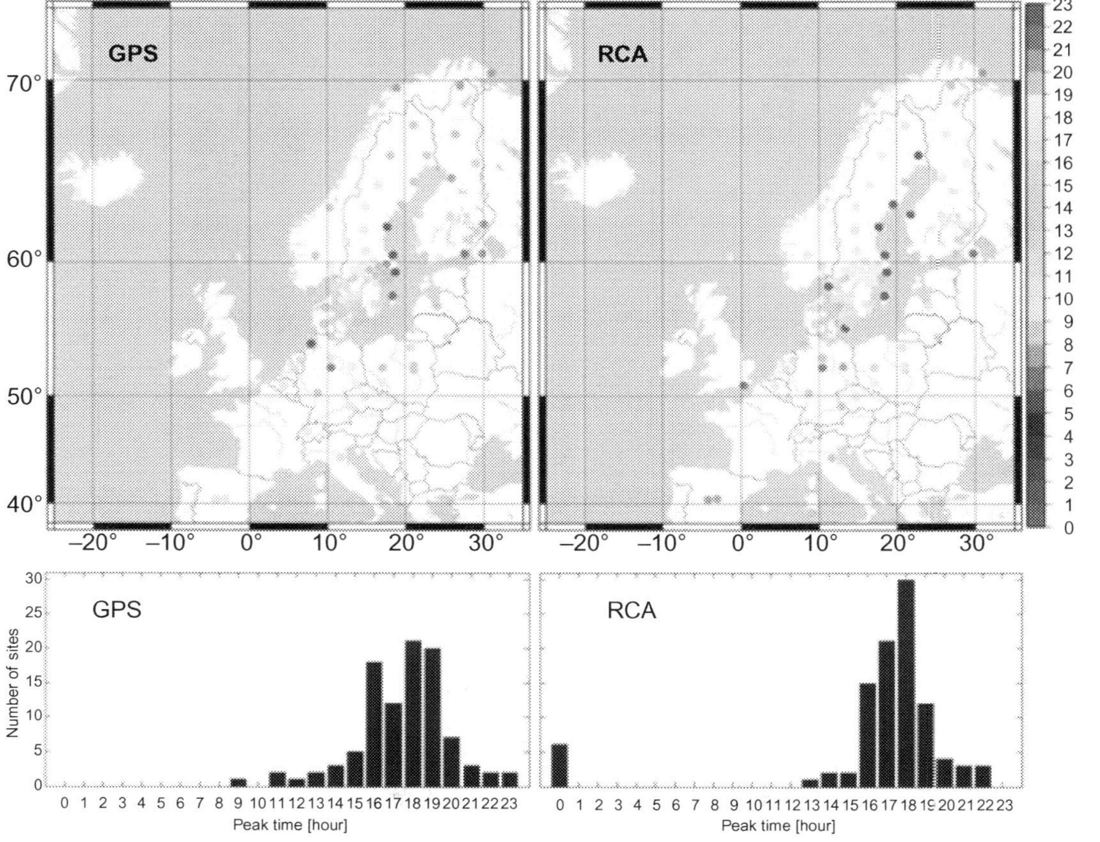

Fig. 7.17 IWV peak time for June–August from: (A) GNSS and (B) RCA. IWV peak time histogram for (C) GNSS and (D) RCA. The time is in local solar time. *(From Ning, T., Elgered, G., Willén, U., & Johansson, J. M. (2013). Evaluation of the atmospheric water vapor content in a regional climate model using ground-based GPS measurements.* Journal of Geophysical Research: Atmospheres, 118*(2), 329–339. https://agupubs.onlinelibrary.wiley.com/doi/full/10.1029/2012JD018053.)*

Table 7.3 List of analyzed GNSS stations from the EPN network.

Site name	LocationIn	EPN since
GOPE	Ondrejov, Czech Republic	31 December 1995
ONSA	Onsala, Sweden	31 December 1995
PENC	Penc, Hungary	03 March 2006
WTZ	RBad Koetzting, Germany	31 December 1995
METS	Kirkkonummi, Finland	31 December 1995

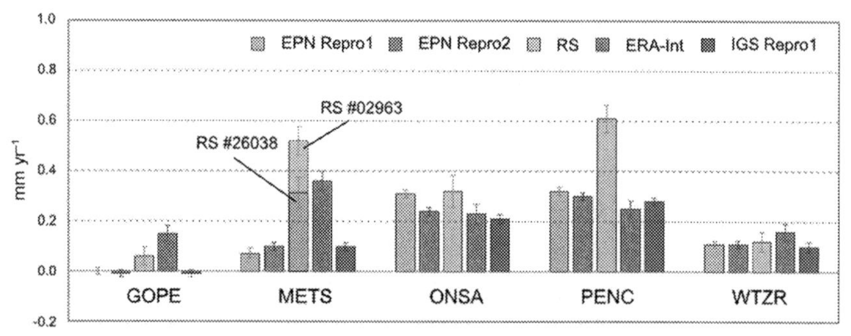

Fig. 7.18 ZTD trends from five GNSS stations from EPN-repro1 *(light green; light gray print version)*, EPN-repro2 *(dark green; dark gray print version)* and IGS-repro1 *(red; dark gray print version)*, radiosonde *(yellow; light gray print version)* and ERA-Interim reanalysis *(blue; dark gray print version). (From Pacione, R., Araszkiewicz, A., Brockmann, E., & Dousa, J. (2017). EPN-Repro2: A reference GNSS tropospheric data set over Europe.* Atmospheric Measurement Techniques, 10*(5), 1689–1705. https://amt.copernicus.org/articles/10/1689/2017/amt-10-1689-2017.pdf.)*

the radiosonde (RMS = 0.16 mm/year). Although the five EPN-repro2 stations examined do not significantly alter the trend detection of ZTD, it has a better agreement with respect to the radiosonde and ERA-Interim data than EPN-repro1. Given the good consistency between trends, EPN-repro2 can be used to detect trends in areas where no other data is available.

GRUAN

One of the difficulties in conducting long-term observations of water vapor is the change in sensors used for aerologic sounding (radiosonde). To improve their accuracy, they are being refined, but this leads to systematic errors and the need for calibration. This makes it difficult to assess humidity changes and to distinguish short-term changes from long-term ones (Wang & Zhang, 2008). Due to the requirements for the long-term sustainability of observations, radiosondes are of limited applicability in climate research (Titchner et al., 2009). In 2006, the Global Climate Observation Network (GCOS/WCRP, n.d.) was added as part of the World Climate Research Program. Within the GCOS two networks for monitoring the upper atmosphere are established—the GCOS Upper-Atmosphere Network (GUAN) and the GCOS Reference Upper-Atmosphere Network (GRUAN, n.d.). The GUAN consists of more than 150 sites, collecting data on a daily basis and employing radiosondes as primary observing methodology. The GRUAN network is much smaller, consisting of more than 30 stations. Unlike GUAN, it uses radiosondes as well as GNSS stations to create a reference network of sensors for monitoring climate change. Measurements from the GRUAN network provide long-term, high-quality climate observations from the Earth's surface, across the troposphere and the stratosphere. The GRUAN data products also include uncertainties of the measurements, providing extensive metadata for each individual measurement. They are used to identify trends, calibrate high-resolution spatial monitoring systems (satellites and aerological sounding), and study atmospheric processes (GCOS134). In 2010, a GRUAN Working Group (GNSS-PW) was set up with the main task of providing high-precision GNSS water vapor measurements to support the study of climate trends and changes.

Sea level monitoring with GNSS

The Tide Gauge Benchmark Monitoring (TIGA) is the lead effort of IGS to define the absolute sea level datum for a set of more than 120 tide gauge stations worldwide. The network was established in 2001 as a working group of IGS (Schöne, Schön, & Thaller, 2009). The tide gauges have been used to measure changes in the ocean level for centuries. A locally installed tide gauge is a relative measurement, which does not take into account the movement of the Earth's crust. Thus GNSS measurements can enhance the tide gauge datum by providing absolute measurements of the vertical movement of the local area landmass. This enables the measurement of absolute sea level changes, rather than the relative change. The TIGA service provides daily coordinates for the GNSS sites, thus enabling the calibration of the tide gauges. These calibrated observations are included in the Global Climate Observing System (GCOS) and are used for calibration of satellite altimeters.

References

Barnett, T. (1978). Estimating variability of surface air temperature in the northern hemisphere. *Monthly Weather Review, 106*(9), 1353–1367.

Bengtsson, L. (2010). The global atmospheric water cycle. *Environmental Research Letters, 5*(2), 025202.

Buehler, S., Von Engeln, A., Brocard, E., John, V. O., Kuhn, T., & Eriksson, P. (2006). Recent developments in the line-by-line modeling of outgoing longwave radiation. *Journal of Quantitative Spectroscopy and Radiative Transfer, 98*(3), 446–457.

Copernicus Program. (n.d.). https://climate.copernicus.eu.

COSMO-REA6. (2021). http://reanalysis.meteo.uni-bonn.de/?COSMO-REA6.

ECMWF Reanalysis. (n.d.). https://www.ecmwf.int/forecasts/datasets/archive-datasets/browse-reanalysis-datasets.

EPN-repro1. (n.d.). https://igs.bkg.bund.de/root_ftp/EPNrepro1/products/.

EPN-repro2. (n.d.). https://igs.bkg.bund.de/root_ftp/EPNrepro2/products/.

ERA-20C. (n.d.). https://www.ecmwf.int/en/forecasts/datasets/reanalysis-datasets/era-20c.

ERA5. (n.d.). https://www.ecmwf.int/en/forecasts/datasets/reanalysis-datasets/era5.

Fujiwara, M., Wright, J. S., Manney, G. L., Gray, L. J., Anstey, J., Birner, T., et al. (2017). Introduction to the SPARC reanalysis Intercomparison project (S-RIP) and overview of the reanalysis systems. *Atmospheric Chemistry and Physics, 17*(2), 1417–1452.

Gattuso, J.-P., Abram, N., & Hock, R. (2019). *Special report on the ocean and cryosphere in a changing climate*. https://www.ipcc.ch/srocc/.

GCOS/WCRP. (n.d.). https://public.wmo.int/en/programmes/global-climate-observing-system.

GRUAN. (n.d.). https://www.gruan.org.

GRUAN reprocessing campaign. (n.d.). http://www.gfz-potsdam.de/en/section/space-geodetic-techniques/projects/gruan.

Hurrell, J. W. (1995). Decadal trends in the North Atlantic oscillation: Regional temperatures and precipitation. *Science, 269*(5224), 676–679.

IGS-repro1. (n.d.). http://acc.igs.org/reprocess.html.

IPCC. (2019). In P. R. Shukla, J. Skea, E. Calvo Buendia, V. Masson-Delmotte, H.-O. Pörtner, D. C. Roberts, & J. Malley (Eds.), *Climate change and land: An IPCC special report on climate change, desertification, land degradation, sustainable land management, food security, and greenhouse gas fluxes in terrestrial ecosystems*. IPCC. In press.

Jones, P. (1989). The influence of ENSO on global temperatures. *Climate Monitor, 17*, 80–89.

Kondratyev, K. Y. (1972). *Radiation processes in the atmosphere*.

Mears, C. A., Santer, B. D., Wentz, F. J., Taylor, K. E., & Wehner, M. F. (2007). Relationship between temperature and precipitable water changes over tropical oceans. *Geophysical Research Letters, 34*(24).

National Snow and Ice Data Center (NSIDC), State of the cryosphere. (2019). https://nsidc.org/cryosphere/sotc/ice_sheets.html.

Ning, T., Elgered, G., Willén, U., & Johansson, J. M. (2013). Evaluation of the atmospheric water vapor content in a regional climate model using ground-based GPS measurements. *Journal of Geophysical Research: Atmospheres, 118*(2), 329–339. https://agupubs.onlinelibrary.wiley.com/doi/full/10.1029/2012JD018053.

Pacione, R., Araszkiewicz, A., Brockmann, E., & Dousa, J. (2017). EPN-Repro2: A reference GNSS tropospheric data set over Europe. *Atmospheric Measurement Techniques, 10*(5), 1689–1705. https://amt.copernicus.org/articles/10/1689/2017/amt-10-1689-2017.pdf.

Parracho, A. C., Bock, O., & Bastin, S. (2018). Global IWV trends and variability in atmospheric reanalyses and GPS observations. *Atmospheric Chemistry and Physics, 18*(22), 16213–16237. https://acp.copernicus.org/articles/18/16213/2018/.

Paxi - The Greenhouse Effect. (n.d.). https://www.esa.int/spaceinvideos/Videos/2018/05/Paxi_-_The_Greenhouse_Effect.

Ross, R. J., & Elliott, W. P. (1996). Tropospheric water vapor climatology and trends over North America: 1973–93. *Journal of Climate*, *9*(12), 3561–3574.

Ross, R. J., & Elliott, W. P. (2001). Radiosonde-based northern hemisphere tropospheric water vapor trends. *Journal of Climate*, *14*(7), 1602–1612.

Schöne, T., Schön, N., & Thaller, D. (2009). IGS tide gauge benchmark monitoring pilot project (TIGA): Scientific benefits. *Journal of Geodesy*, *83*(3), 249–261.

Solomon, S., Manning, M., Marquis, M., & Qin, D. (2007). *Climate change 2007-the physical science basis: Working group I contribution to the fourth assessment report of the IPCC. Vol. 4.* Cambridge University Press.

TIGA reprocessing campaign. (n.d.). http://adsc.gfz-potsdam.de/tiga/index_TIGA.html.

Titchner, H. A., Thorne, P. W., McCarthy, M. P., Tett, S. F., Haimberger, L., & Parker, D. E. (2009). Critically reassessing tropospheric temperature trends from radiosondes using realistic validation experiments. *Journal of Climate*, *22*(3), 465–485.

Trenberth, K. E., Dai, A., Rasmussen, R. M., & Parsons, D. B. (2003). The changing character of precipitation. *Bulletin of the American Meteorological Society*, *84*(9), 1205–1218.

Trenberth, K. E., Smith, L., Qian, T., Dai, A., & Fasullo, J. (2007). Estimates of the global water budget and its annual cycle using observational and model data. *Journal of Hydrometeorology*, *8*(4), 758–769.

Vey, S., Dietrich, R., Rülke, A., Fritsche, M., Steigenberger, P., & Rothacher, M. (2010). Validation of precipitable water vapor within the NCEP/DOE reanalysis using global GPS observations from one decade. *Journal of Climate*, *23*(7), 1675–1695.

Wallace, J. M., Zhang, Y., & Renwick, J. A. (1995). Dynamic contribution to hemispheric mean temperature trends. *Science*, *270*(5237), 780–783.

Wang, J., & Zhang, L. (2008). Systematic errors in global radiosonde precipitable water data from comparisons with ground-based GPS measurements. *Journal of Climate*, *21*(10), 2218–2238.

Further reading

Halpert, M. S., & Ropelewski, C. F. (1992). Surface temperature patterns associated with the southern oscillation. *Journal of Climate*, *5*(6), 577–593.

Heald, C. L., Ridley, D. A., Kroll, J. H., Barrett, S. R. H., Cady-Pereira, K. E., Alvarado, M. J., et al. (2014). Contrasting the direct radiative effect and direct radiative forcing of aerosols. *Atmospheric Chemistry and Physics*, *14*, 5513–5527. https://doi.org/10.5194/acp-14-5513-2014.

Wang, B., & Wang, Y. (1996). Temporal structure of the southern oscillation as revealed by waveform and wavelet analysis. *Journal of Climate*, *9*(7), 1586–1598.

CHAPTER 8

GNSS reflectometry (GNSS-R) for environmental observation

The foundations behind the utilization of reflected GNSS signals for environmental research were laid down in the 1990s. Several pioneering works laid the groundwork for using these "signals of opportunity." Martin-Neira proposed the Passive Reflectometry and Interferometry System (PARIS) for using signals of opportunity for ocean altimetry (Martin-Neira, 1993). The developed methodologies can be classified into categories, based on platforms and signals used. The platform classification includes spaceborne, airborne, and ground-based techniques. Classification based on the signal property includes Doppler-delay map (DDM), signal-to-noise ratio (SNR), and carrier-phase observation techniques. In terms of receiving station's setups, there are upward-looking and side-looking single-antenna setups, as well as dual-antenna setups, i.e., one upward-looking and one side- or downward-looking antenna. Each technique is briefly described in this chapter. These techniques enable the measurement of (1) wind speed over water bodies (see "GNSS-R wind speed measurements" section), (2) water level height including sea level (see "GNSS-R water level/land height measurements" section) and land elevation, (3) soil moisture ("GNSS-R soil moisture monitoring" section), (4) vegetation, and (5) snow height observations (see Fig. 8.1).

Spaceborne GNSS Reflectometry resembles the Synthetic-Aperture Radar (SAR) methodology of probing the Earth's surface. In SAR systems the transmitter and receiver are situated on the same satellite, orbiting the Earth, and use the same antenna (these systems are known as monostatic radar systems). The SAR satellites usually operate in LEO and, in general, provide data on the distance traveled by the signal from the transmitter, reflected by the Earth's surface and back to the onboard receiver. The GNSS-R satellite measurements can be categorized as multistatic radars. Conceptually, the signal transmitted from the MEO GNSS satellites is reflected from the Earth's surface and then received by an LEO satellite. Unlike SAR, each GNSS-R LEO satellite can observe reflections from

Global Navigation Satellite System Monitoring of the Atmosphere
https://doi.org/10.1016/B978-0-12-819152-1.00002-7

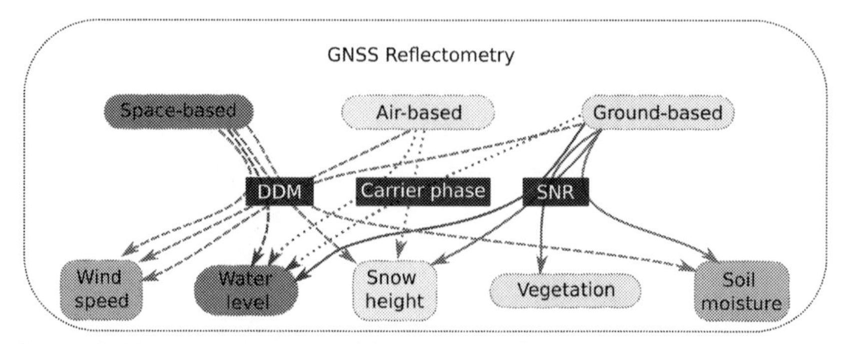

Fig. 8.1 GNSS overview. Overview of the GNSS-R platforms, methods, and applications. The *arrows* connect each platform with the measured environmental property through the used methodology (solid lines indicate SNR, dashed lines—DDM's, and dotted—carrier phase methods).

multiple MEO GNSS satellites simultaneously, but the receiving satellite has no control over the direction from which the reflections are coming.

Theoretical background—Polarization

When an electromagnetic wave is reflected from a surface, the wave interacts with the reflective surface in accordance with Brewster's law. Brewster's law states that an unpolarized electromagnetic wave reflected above a certain threshold incidence angle is known as Brewster angle (Fig. 8.2).

$$\theta = \arctan\left(\frac{n1}{n2}\right), \tag{8.1}$$

The electromagnetic wave polarized in the same plane as the incident electromagnetic wave and the surface normal at the point of incidence is not reflected.

When the right-hand circular polarized (RHCP, see Chapter 2) GNSS signals are reflected from the surface of the Earth, they are subject to Fresnel's reflectivity relationship

$$E_p = R_{pq} E_q, \tag{8.2}$$

where E_p is the electric field of the scattered signal with polarization p, E_q is the incidence electric field with polarization q, and R_{pq} is the Fresnel coefficient for q- to p-polarized reflection. This coefficient depends on the dielectric properties of the reflective surface, as well as on the incident angle. For each polarization pair, the R coefficient is different. The sum of the R coefficients for any reflection is $R < 1$, since the reflection cannot generate

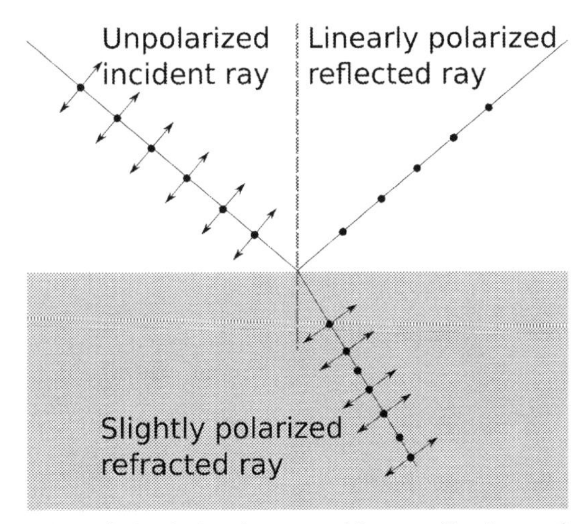

Fig. 8.2 Brewster. Unpolarized signal, approaching a reflective surface at incidence angles larger than the Brewster angle, is polarized at reflection with linear polarization, perpendicular to the plane, defined by the incoming signal and the normal of the reflective surface. *(Reproduced from https://en.wikipedia.org/wiki/Brewster%27s_angle#/media/File:Brewsters-angle.svg.)*

an electric field stronger than the incidence (Conservation law). Eq. (8.2) shows that part of the reflected electromagnetic wave changes its polarization upon reflection, while part of it maintains its polarization.

The portion of the reflected GNSS signal, which maintains its RHCP polarization, depends on the Brewster law. When the reflected GNSS signal is coming from close to nadir, most of the reflected signal will change its polarization to LHCP. At the Brewster angle, the power of the reflected RHCP and LHCP is equal. Since the Brewster angle depends on the dielectric properties of the surface, it has to be measured for each reflective surface (see Fig. 8.3). Different phases of water have Brewster angles in the interval between 54 degree for dry snow (blue curves) and 83 degree for seawater. The Brewster angle for seawater can itself vary between 58 degree and 85 degree, depending on water salinity and density (Jin, Cardellach, & Xie, 2014).

The standard GNSS antennas and receivers, used for positioning and geodesy, are capable of receiving RHCP signals only. Dedicated LHCP antennas and receivers are developed specifically to observe reflected signals. Such receivers are not widely available and are primarily used in the scientific ground-based stations (described in "Ground-based GNSS-R measurements" section), or on satellites and aircraft.

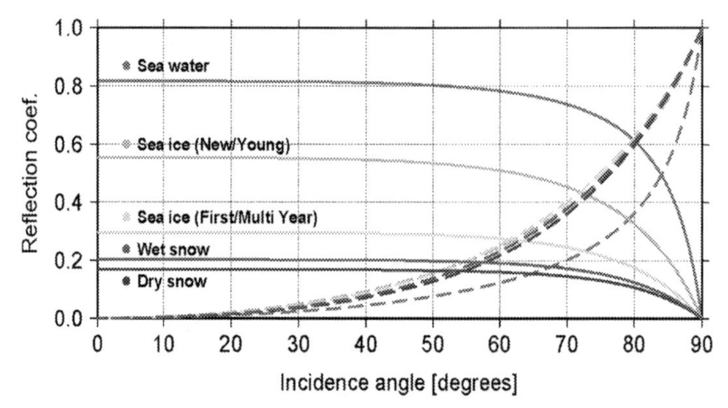

Fig. 8.3 The *solid lines* show the reflection coefficient R of the LHCP reflected signal at incidence angles between 0 degree and 90 degree for seawater, ice, and snow. The *dashed lines* show the reflection coefficient for the RHCP reflected signal. The Brewster angles for each reflecting surface are at the intersection between the *solid* and *dashed lines.* This figure shows that different phases of water have Brewster angles in the interval between 54 degree for dry snow (blue curves; dark gray in print version) and 83 degree for seawater. *(From Jin, S., Cardellach, E., & Xie, F. (2014). GNSS remote sensing. Springer.)*

GNSS-R wind speed measurements

Specular and diffuse scattering

The theoretical basis for GPS signals scattering from the ocean surface through the Bistatic Radar Equation was proposed (Zavorotny & Voronovich, 2000). The roughness of the reflective surface plays an important role in the way a receiver reads the reflected signals. In an ideal case, when the reflective surface is smooth, the reflected signal received by the observer originates in a single point (specular reflection). In reality, the surface of the Earth is rough and the reflected signal received by the observer originates from a glistening zone (Fig. 8.4). The size of the zone depends on the roughness of the surface.

When the GNSS signal is reflected from a smooth surface, the receiver observes two navigation messages—one coming directly from the satellite and the other from the reflective surface. The excess path of the reflected signal is determined through the cross correlation between the received direct signal binary code from the satellite and the binary code of the reflected signal. Since the reflection is specular, the waveform of the correlator output for the reflected signal is the same as the waveform for the direct signal (see Fig. 8.5) and has triangular function of half-width equal to the travel time of a single code chip.

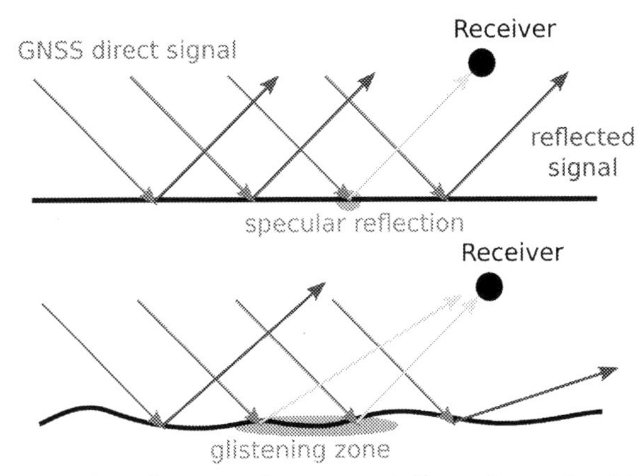

Fig. 8.4 Glistening. The reflective surface on top is flat and smooth, which means that the observer receives reflection from a single point. When the roughness of the reflective surface increases, the observer receives reflected signals from multiple points within a glistening zone. The higher the roughness, the larger the glistening zone.

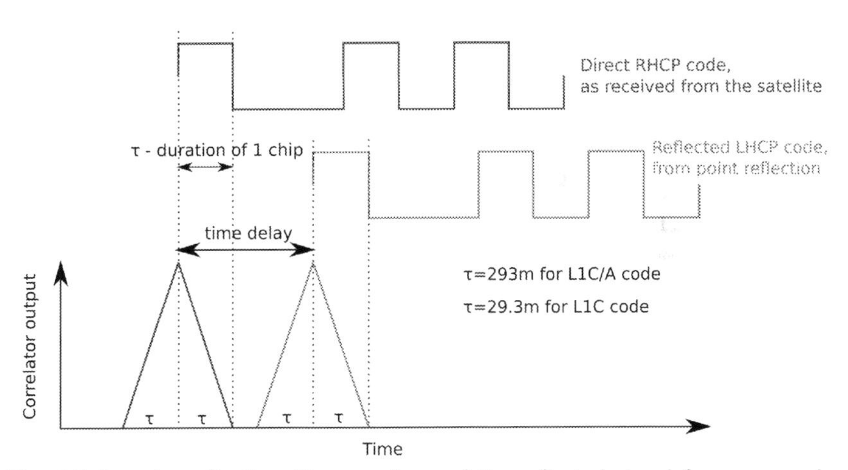

Fig. 8.5 Specular reflection. The waveform of the reflected signal from a specular reflection is identical to the waveform of the direct signal. The width of the waveform is double the time duration of one code chip.

When the GNSS signal is reflected from a rough surface, the signal travels a distance different from the satellite to the receiver than when it is reflected from the center facets of the glistening zone, compared to the reflections, coming from the facets at the edges of the glistening zone. Since the signal is traveling at a constant speed, the longer travel distance is observed by the

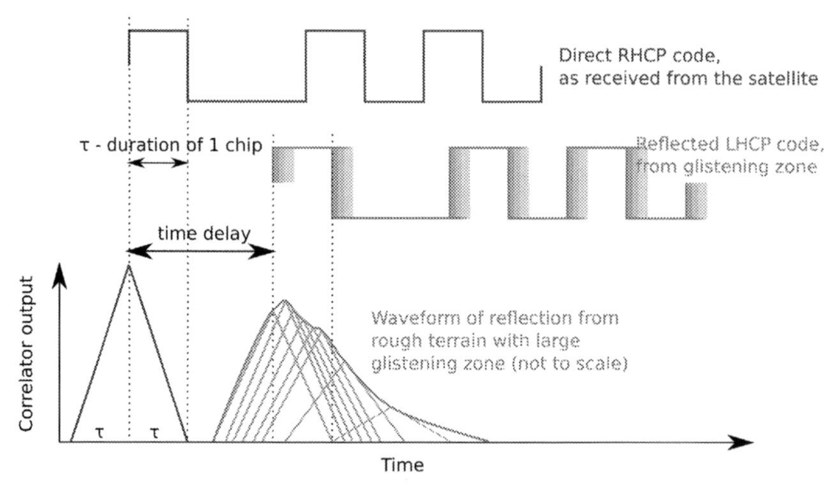

Fig. 8.6 Glistening waveform. The waveform of a reflected signal from a rough surface is prolonged since reflections from multiple facets within the glistening zone are received. The larger the glistening zone, the wider the waveform.

receiver as a delay. Each reflection facet contributes its own delayed copy of the original signal code. All of these copies are received in the receiver with incremental delays. These delays contribute to a prolonged waveform, demonstrating the longer time that the reflected signals take to reach the receiver. The higher the roughness of the reflective surface, the flatter the waveform (see Fig. 8.6).

Since the GNSS signals are transmitted from a moving satellite, reflected from a quasistationary surface and received on a moving platform, Doppler shifts in the frequency of the received signals are observed. A separate waveform can be observed for each Doppler belt, thus creating a Doppler-delay map (DDM) of the reflection (see Fig. 8.7).

Spaceborne wind speed measurements

The scientific principle employed in wind speed measurements from dedicated LEO satellites, equipped with GNSS-R receivers, is based on the increased roughness of the sea surface. The stronger the blowing wind, the higher the surface roughness. In 2002 the first reflected GPS signal was received from space through the Spaceborne Imaging Radar-C (SIR-C) (Lowe et al., 2002). The first satellite, equipped with a GNSS-R receiver, is the UK Disaster Monitoring Constellation (UK-DMC), which

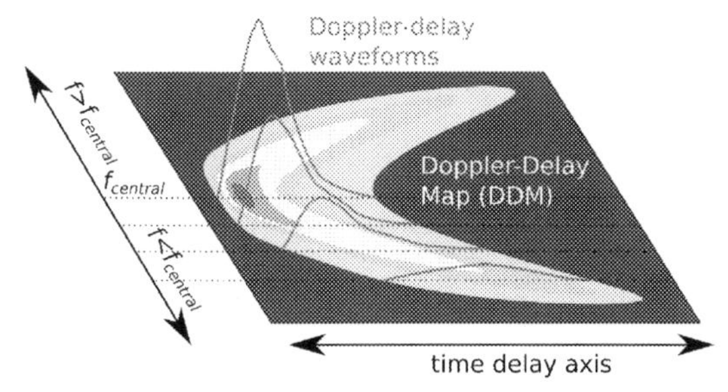

Fig. 8.7 DDM. The Doppler-delay maps are visual representations of the waveforms of a reflection, stacked along the frequency Doppler axis (y axis) and are color coded for intensity. The higher the roughness of the reflective surface, the larger the footprint of the reflection on the DDM.

was launched in 2003. During this mission, the first wind speed measurements were derived through GNSS-R measurements (Gleason et al., 2005). In 2014 the first dedicated GNSS-R mission was launched—the TechDemoSat-1 (TDS-1). The TDS-1 mission was followed in 2016 by the eight satellite Cyclone GNSS (CYGNSS) mission for tropical cyclone forecasting and hurricane tracking (Ruf et al., 2018).

The method of measuring wind speed from DDMs (Fig. 8.8) relies on creating a Geophysical Mapping Function (GMF). The Geophysical Mapping Function should not be confused with Global Mapping Function (also marked as GMF), which is used in positioning. The GMF is a function, designed to relate the waveforms from the DDMs to wind speed observations. The training of these generally empirical GMFs is based on using supplementary data from global reanalysis (such as ERA5 or ERA-Interim), or satellite measurements from missions, such as the Advanced SCATterometer (ASCAT) (Asgarimehr, Zavorotny, Wickert, & Reich, 2018). In figure GMF, DDM cross sections from the TDS-1 satellite are plotted against 10-m wind speeds from ERA-Interim. The red line (dark gray in print version) indicates the fitted exponential GMF. The indicated bias between the ERA-Interim and GNSS-R wind speeds is 0.23 m/s with RMSE of 2.76 m/s (Fig. 8.9).

The experiments with measuring wind speed using GNSS-R have shown absolute biases between GNSS-R measurements and ASCAT in the range of 0–1.5 m/s and RMSE of 1–4 m/s. The exponential form of the GMFs suggests that the GNSS-R observations of wind speed are far more sensitive at wind speeds below 15 m/s, than at higher wind speeds.

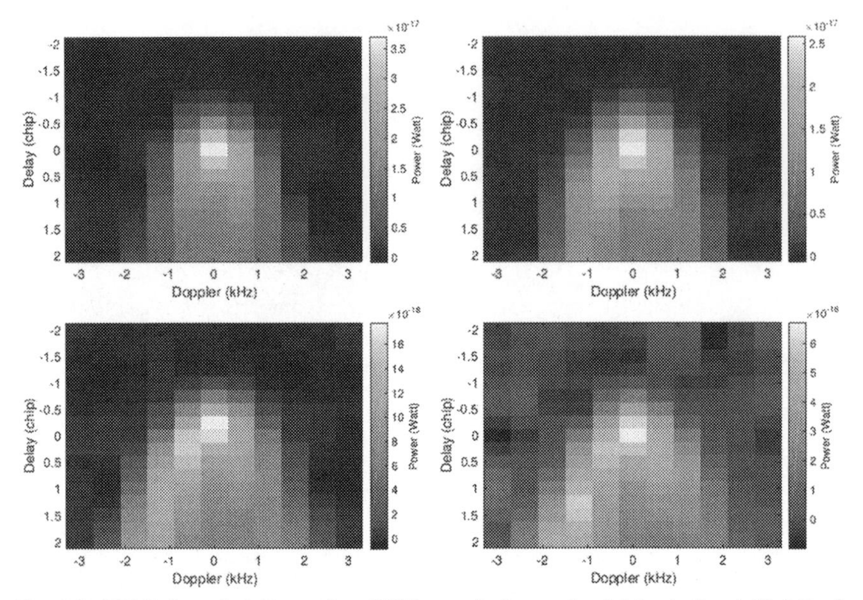

Fig. 8.8 DDM's for wind. Exemplary DDMs at wind speeds of 1.5 m/s *(top left)*, 3.7 m/s *(top right)*, 4.5 m/s *(bottom left)*, and 8.3 m/s *(bottom right)*. *(From Asgarimehr, M. (2020). Spaceborne GNSS Reflectometry: Remote Sensing of Ocean and Atmosphere. In Technische Universität Berlin. Technische Universität Berlin.)*

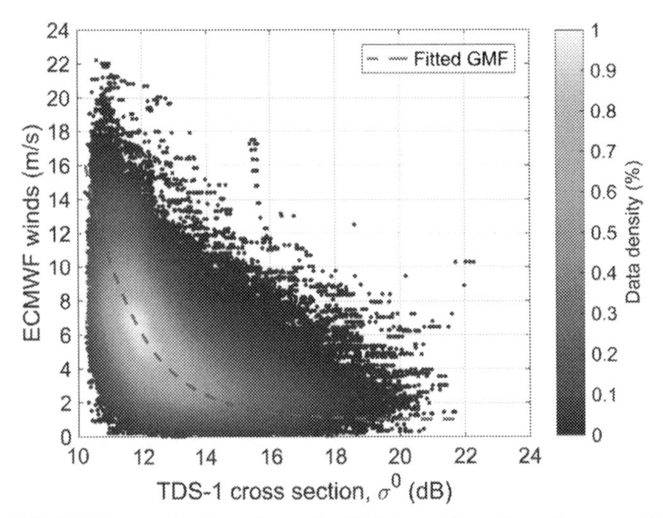

Fig. 8.9 GMF. DDM cross sections from the TDS-1 satellite plotted against 10-m wind speeds from ERA-Interim. The *red line* (dark gray in print version) indicates the fitted exponential GMF. *(From Asgarimehr, M., Wickert, J., & Reich, S. (2018). TDS-1 GNSS reflectometry: Development and validation of forward scattering winds. IEEE Journal of Selected Topics in Applied Earth Observations and Remote Sensing, 11(11), 4534–4541.)*

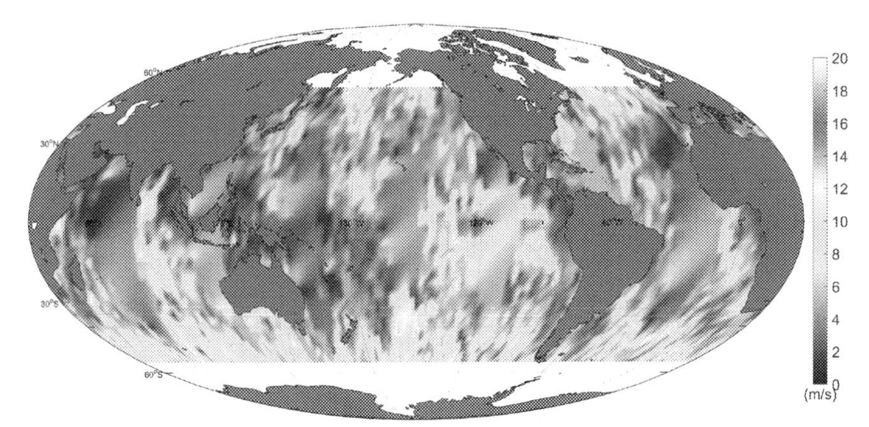

Fig. 8.10 Wind world. Average global wind speed (May 2015 to June 2016) by TDS-1 SGR-ReSI measurements. *(From Asgarimehr, M., Wickert, J., & Reich, S. (2018). TDS-1 GNSS reflectometry: Development and validation of forward scattering winds. IEEE Journal of Selected Topics in Applied Earth Observations and Remote Sensing, 11(11), 4534–4541.)*

Additionally, the sea roughness is affected not only by the blowing wind but also by the splashes of raindrops on the sea surface, which brings additional uncertainty to the final wind speed measurements (Asgarimehr, Wickert, & Reich, 2018). The greatest advantage of the GNSS-R measured wind speeds is the high spatiotemporal resolution of measurements, provided by the growing constellation of GNSS-R satellites (see Fig. 8.10).

GNSS-R water level/land height measurements

Carrier-phase measurements

Phase altimetry relies on measuring distance differences based on phase change and the number of full phase cycles, each of them representing a distance change of 1 wavelength

$$\Delta\phi = \frac{1}{\lambda} p_H + \upsilon, \tag{8.3}$$

where $\Delta\phi$ is the phase change, λ is the wavelength, p_H is the path for a specific height, and υ are residuals. By having two receivers, one RHCP upward-looking and another LHCP downward- or side-looking, the difference in height between these two receivers can be calculated through the path difference between the two receivers. The change in the path difference can be monitored through the change in the carrier phase.

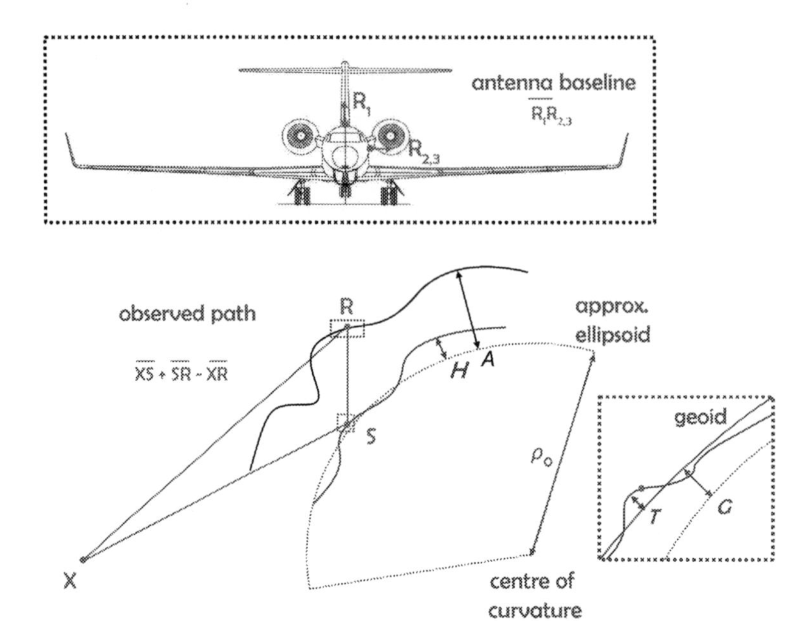

Fig. 8.11 HALO. This figure shows the location of the zenith-looking and side-looking GNSS antenna onboard the HALO aircraft during a measurement campaign over Italy. Below the geometry of the GNSS signals is presented. *(Reproduced from Semmling, A., Beckheinrich, J., Wickert, J., Beyerle, G., Schön, S., Fabra, F., Pflug, H., He, K., Schwabe, J., & Scheinert, M. (2014). Sea surface topography retrieved from GNSS reflectometry phase data of the GEOHALO flight mission. Geophysical Research Letters, 41(3), 954–960.)*

Sea level measurements with airborne GNSS-R

Airborne GNSS Reflectometry uses the carrier-phase GNSS-R. Unlike the spaceborne systems, the receivers are mounted on an aircraft, such as the High Altitude and Long Range Research Aircraft (HALO), with two antennas, one looking in zenith and the second one in nadir direction (Fig. 8.11).

This method is used for observation campaigns, where reflections from specific areas are of interest. Such campaigns have been carried out for water level measurements over the lake Konstanz (Semmling, Beckheinrich, et al., 2014) and sea level and wind speed measurements over the southern coast of Italy (see Fig. 8.12) (Semmling, Schön, et al., 2014). These campaigns have been used both for measurements over specific regions and for proof-of-concept missions for GNSS-R.

The experiment over the coast of Italy shows very good agreement between the GNSS-R-derived sea altimetry data and the Mean Sea Surface height from the DTU-10 ellipsoid observations (Fig. 8.13). The reported accuracy for some of the measured tracks in the figure map of Italy is below 10 cm.

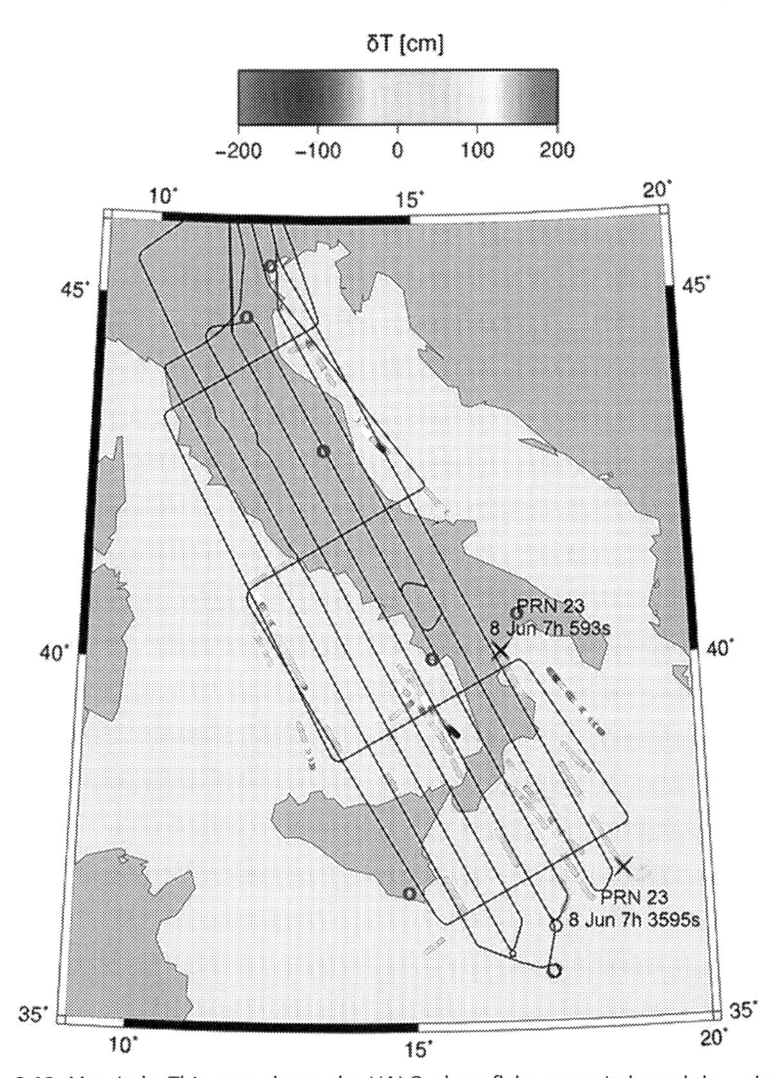

Fig. 8.12 Map Italy. This map shows the HALO plane flybys over Italy and the colored tracks represent the observed GPS reflections from the sea surface. *(From Semmling, A., Beckheinrich, J., Wickert, J., Beyerle, G., Schön, S., Fabra, F., Pflug, H., He, K., Schwabe, J., & Scheinert, M. (2014). Sea surface topography retrieved from GNSS reflectometry phase data of the GEOHALO flight mission.* Geophysical Research Letters, 41*(3), 954–960.)*

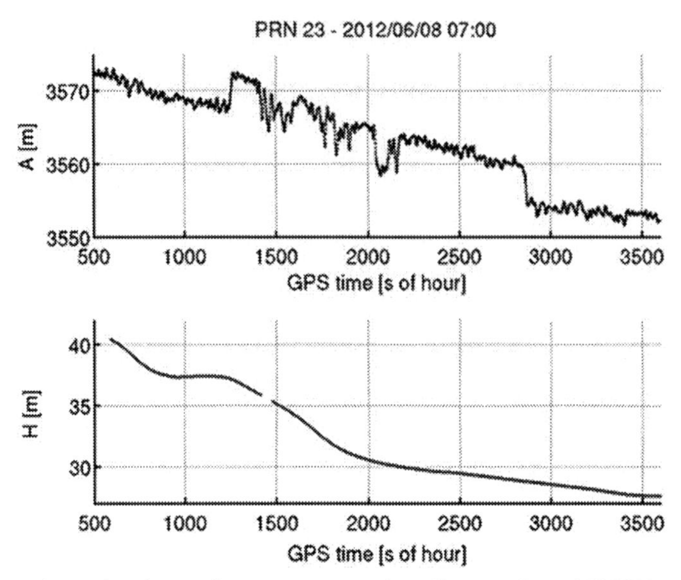

Fig. 8.13 Italy results. The *top* figure represents the reflections from GPS PRN23 satellite. The *bottom* figure is the mean sea surface height over the same path, as the PRN23 reflection. *(From Semmling, A., Beckheinrich, J., Wickert, J., Beyerle, G., Schön, S., Fabra, F., Pflug, H., He, K., Schwabe, J., & Scheinert, M. (2014). Sea surface topography retrieved from GNSS reflectometry phase data of the GEOHALO flight mission. Geophysical Research Letters, 41(3), 954–960.)*

The retrievals are carried out at elevation angles between 10° and 30° since the method postulates specular reflections and higher elevation angles are more dependent on sea roughness (Semmling, Beckheinrich, et al., 2014).

Ground-based GNSS-R measurements

Apart from airborne GNSS-R observations can be also performed with ground-based GNSS stations. Experiments in the Onsala Observatory in Sweden proved the concept of using a zenith-looking RHCP antenna, parallel to a nadir-looking LHCP antenna to perform carrier-phase observations, similar to the airborne altimetry methodology. In the same study, the RHCP antenna was also used to record SNR data (described in more detail in "GNSS-R soil moisture monitoring" section) to measure the height of the reflective surface. The results from these experiments are then compared to the tide gauge, installed near the GNSS setup (Fig. 8.14).

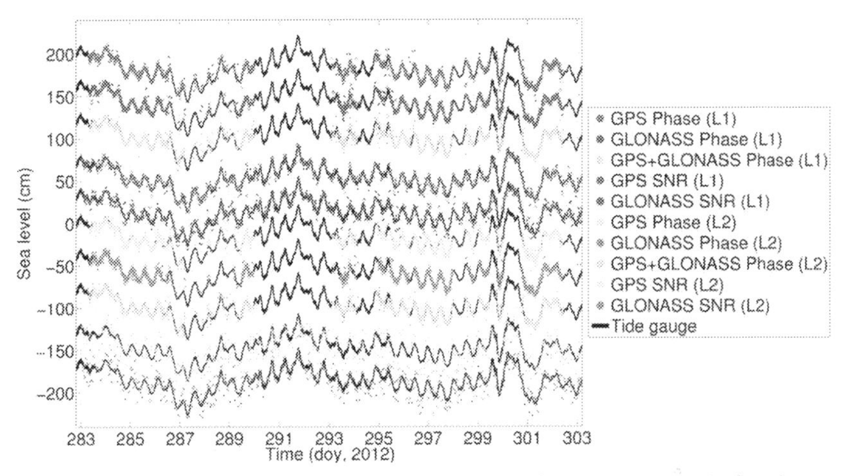

Fig. 8.14 Onsala. Sea level height from tide gauge *(in black)* over a period of 20 days at Onsala Observatory are compared with GNSS reflectometry observations *(in color)* from several GNSS combinations and two reflector height retrieval techniques—carrier phase and SNR. *(Löfgren, J. S., & Haas, R. (2014). Sea level observations using multisystem GNSS reflectometry. NKG, 17th General Assembly.)*

The results from this experiment show (see Fig. 8.14) that both carrier-phase observations from the dual setup and SNR observations from the RHCP zenith antenna provide accurate measurements of the tides in the area of observation (Löfgren & Haas, 2014). The correlation between the GNSS-R measurements and the tide gauge is between 0.85 and 0.95 for the different methods and combinations of different GNSSs. The study also shows that the SNR observations have 1.5–3 times higher standard deviation from the tide gauge compared to the results from the carrier-phase observations.

GNSS-R soil moisture monitoring

The soil can be separated into several different layers, based on porosity, presence of organic matter, and structure. The layers in the soil are named horizons (Fig. 8.12). The soil layer, located in the top 20 cm of the soil profile is the O horizon (O stands for Organic). The O horizon is rich in organic particles, holds the root structure of the low vegetation, and is very porous. The following A (20–50 cm) and B (50 cm–3 m) horizons are denser, less porous, and host the roots of trees. The C horizon (3–5 m) is the soil layer just above the bedrock. This is usually the layer where groundwater is found. When speaking about soil moisture, hydrologists usually consider the amount of water in the O, A, and B horizons. GNSS signals, however, get reflected from the O horizon, so the soil moisture, measured through GNSS-R, is only relevant to the topsoil, i.e., top 15 cm (Brady & Weil, 2013).

The most widely used soil moisture metrics are (1) specific water content (SWC), which is the fraction of the water mass from the mass of a confined amount of soil, and (2) volumetric water content (VWC), which is the fraction of the water volume in a confined volume of soil. VWC can be measured in both $[cm^3_{H2O}/cm^3_{soil}]$ or in $[Vol\%]$, where the absolute values have the following relation: $1[cm^3_{H2O}/cm^3_{soil}] = 100[Vol\%]$. Another soil moisture metric is the relative water content (RWC), which represents the volume of water present in the soil as a fraction of the saturated water amount in the soil. RWC is measured in [%] (Simeonov, 2021).

Single-antenna SNR observations

When the direct and reflected GNSS signals reach the ground antenna, they are both detected. The signal strength of the direct and the reflected signal, as received by the antenna, depends on the elevation angle of the GNSS satellite (through the Brewster law dependence) and on the GNSS antenna gain pattern (Larson et al., 2008). Since reflected signals are usually regarded as noise, geodetic antennas are equipped with noise-canceling choke rings (specially designed metal collars, which block the reflected signals). The antenna cannot distinguish the direct signal from the reflected, the recorded signal strength shows the interference pattern between these two signals (as shown in Fig. 8.15). The reason for the occurrence of the interference pattern is that there is a periodicity in the phase difference between the direct and reflected signals (Fig. 8.16). When the direct and reflected signals are in-phase, the resulting amplitude of the received signal strength is amplified. When the two signals are in the counter phase, the signal strength, as received by the antenna, is dampened. The term signal strength is used as a description of the direct signal in units of dB—Hz.

The resulting signal strength interference pattern is polynomially detrended and subsequently, the low elevation angles of the signal-to-noise ratio (SNR) is acquired and analyzed. The SNR can be modeled with the following equation:

$$SNR = A\cos\left(\frac{4\pi h}{\lambda} + \phi\right) \tag{8.4}$$

where A is the average amplitude of the interference pattern, h is the height difference between the antenna and the reflective surface, λ is the frequency

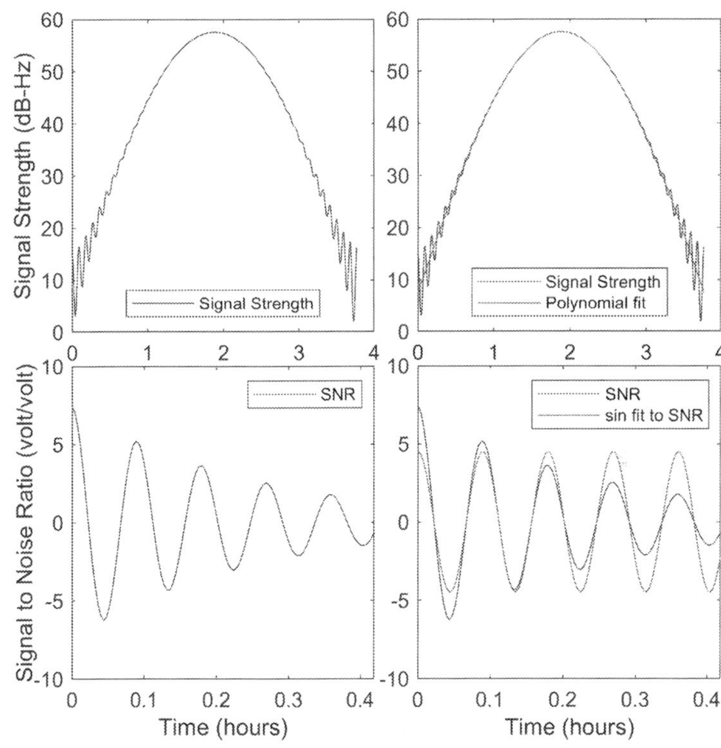

Fig. 8.15 Single antenna reflection. *Top left* is the theoretical signal strength, observed by a ground-based station for a GPS satellite transit. A second-order polynomial fit is used to simulate the signal strength observations without the reflection interference *(top right)*. The difference between the observed and the polynomial fit represents the SNR interference pattern *(bottom left)*, the period and amplitude of which are estimated through an LSA sinusoidal fit *(bottom right)*. *(From Simeonov, T. (2021). Derivation and analysis of atmospheric water vapor and soil moisture from ground-based GNSS stations. In Technische Universität Berlin. Technische Universität Berlin.)*

of the signal, e is the elevation angle of the satellite, ϕ is a phase shift, and the SNR is the result of the detrending of the signal strength and is measured in (volt/volt). Fig. 8.15 (bottom left and right) shows that the amplitude depends on the elevation angle.

The amplitude and phase shift of the SNR are estimated using a Lomb-Scargle least-squares adjustment (LSA). Every sampled frequency is represented in the spectrum of significance, where the highest significance is given to the lowest least-squares difference between the sample and the fitted curve.

This SNR methodology enables the measurement of reflector height changes (used in "Ground-based GNSS-R measurements" section), as well as soil moisture observations through the changes of the phase ϕ.

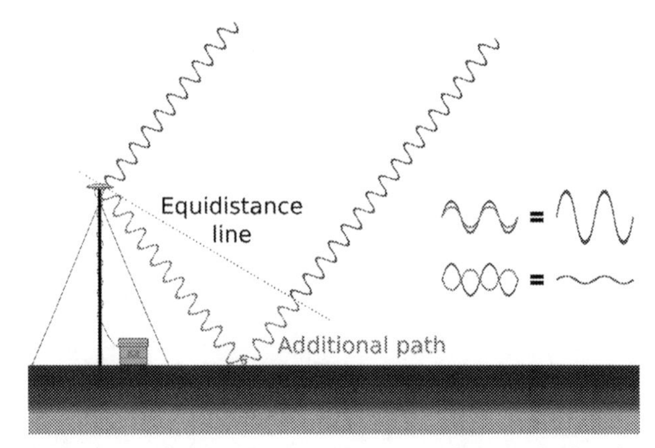

Fig. 8.16 SNR distance. The GNSS antenna cannot distinguish the direct from the reflected RHCP signal. Thus the antenna records the interference between the direct and reflected signals in the signal strength observation. *(From Simeonov, T. (2021). Derivation and analysis of atmospheric water vapor and soil moisture from ground-based GNSS stations. In Technische Universität Berlin. Technische Universität Berlin.)*

The GNSS signal path is usually described as a straight line (direct optical path). In reality, the signal propagation between the GNSS satellite and the receiver forms an electromagnetic wave front (Fig. 8.17). The prolate spheroid volume, encapsulating semicoherent signals with phase shifts up to 90 degree from the direct optical path, is the first Fresnel zone (visualized in Fig. 8.17). The section of the first Fresnel zone of a reflected signal has an elliptical shape (see Fig. 8.18).

SNR soil moisture monitoring with single-antenna GNSS-R

The soil moisture estimations are extracted from the phase shifts of the SNR data. (Chew, Small, Larson, & Zavorotny, 2014) estimated that the change of 0.65° in the phase shift of the SNR is equivalent to a 1 $Vol\%$ change of soil moisture. Since the relation between the two parameters is linear, a dataset with the phase changes can be created

$$\Delta\phi = \phi - \phi_0 \tag{8.5}$$

where ϕ_0 is the minimum phase shift of the SNR for the entire dataset. From this the volumetric water content (VWC) of the soils can be estimated as follows:

$$VWC = \frac{\Delta\phi}{\gamma} + c_{min} \tag{8.6}$$

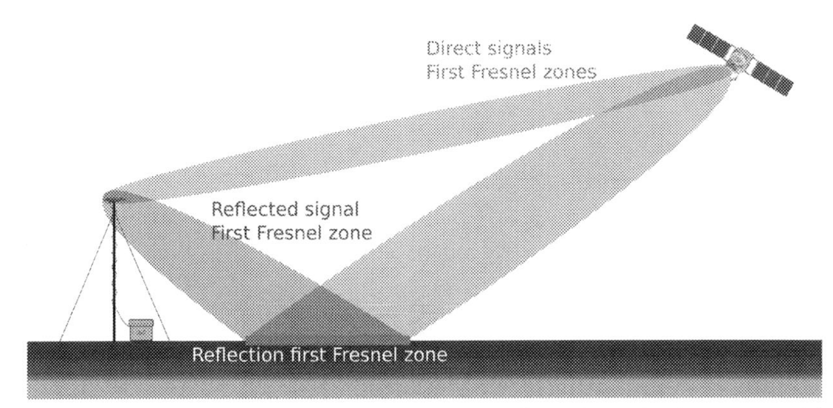

Fig. 8.17 Fresnel. First Fresnel zones of direct and reflected signals. The section of the reflected first Fresnel zone indicates the area of signal reflection at the surface. This is not to be confused with the reflection glistening zone, discussed in "Theoretical background—Polarization" section. *(Reproduced from Simeonov, T. (2021). Derivation and analysis of atmospheric water vapor and soil moisture from ground-based GNSS stations. In Technische Universität Berlin. Technische Universität Berlin.)*

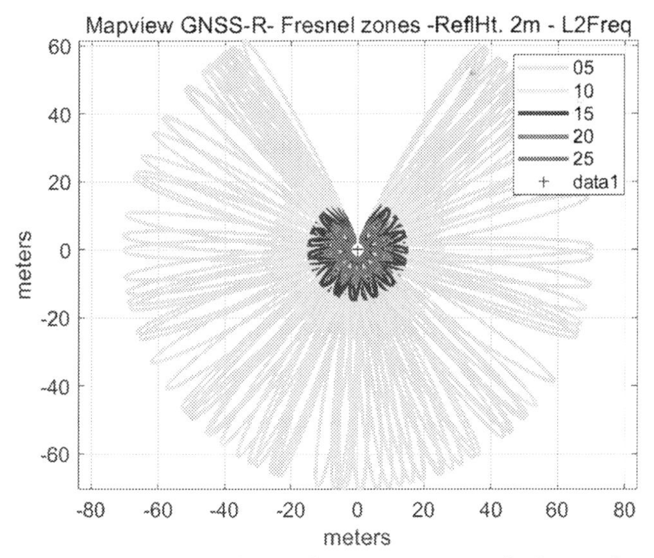

Fig. 8.18 Fresnel ground. This is a figure of the Fresnel zones of reflection for a 2-m high antenna above ground *(marked with a cross in the center)* and for elevation angles between 5 degree *(green ellipses; dark gray in print version)* and 25 degree *(red ellipses; dark gray in print version)*. The area covered by these reflections is about 500 m^2.

where c_{min} is a constant, equal to the minimum soil moisture or residual soil moisture. The value of c_{min} is station specific and is within the range 3.5–5 Vol%. Apart from a minimum possible value for soil moisture, there is also a saturation soil moisture value. The saturation soil moisture is estimated to be 60 ± 10 Vol%, depending on the soil type (Brady & Weil, 2013).

A study at an experimental site in Germany (Marquardt) shows that the soil moisture retrievals from a ground-based GNSS station are comparable to soil moisture measurements performed by Time Domain Reflectometry (TDR). The correlation between the two methods shows variability between the different seasons and is in the range between 0.5 and 0.9 with higher correlations in the summer periods and lower in winter. The seasonality is explained by several factors, among which the freezing of the soil during cold winters with no snow cover. It is to be noted that there is high correlation between precipitation events (blue bars) and local maxima of soil moisture. The TDR measurements during late August and September (Fig. 8.19) show much lower soil

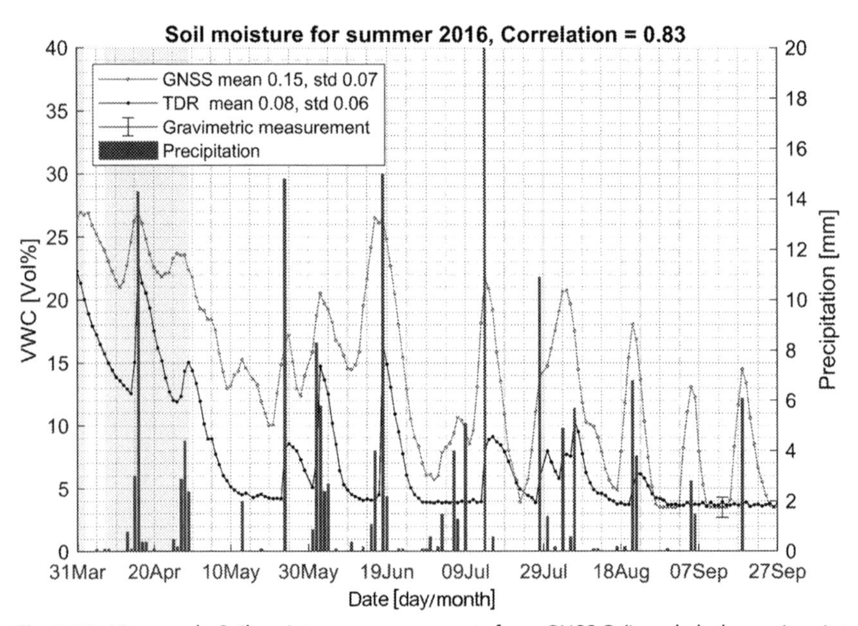

Fig. 8.19 Marquardt. Soil moisture measurements from GNSS-R *(in red; dark gray in print version)* and TDR *(in black)* for the summer months of 2016 at GNSS station Marquardt, Germany. Precipitation is plotted in *blue bars* (dark gray in print version). *(From Simeonov, T. (2021). Derivation and analysis of atmospheric water vapor and soil moisture from ground-based GNSS stations. In Technische Universität Berlin. Technische Universität Berlin.)*

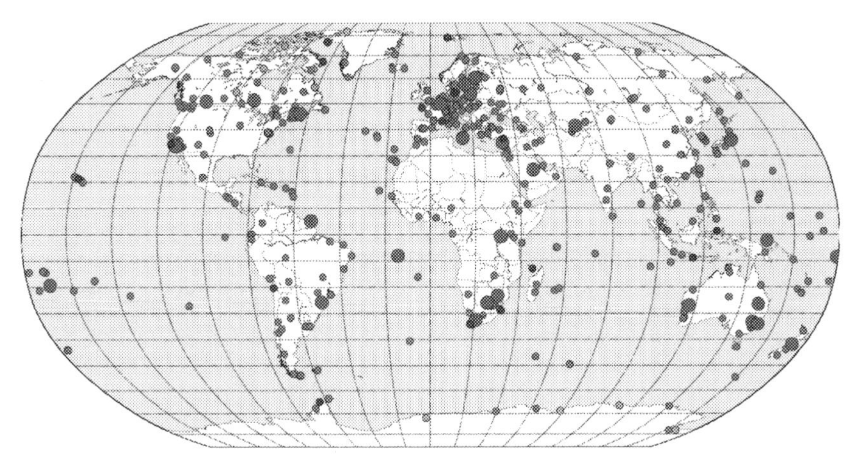

Fig. 8.20 IGS. The stations, marked with the large *green dots* (large dark gray in print version) indicate sites, where GNSS-R soil moisture observations have been made. The sites marked with *blue* (dark gray in print version) and *cyan* (light gray in print version) indicate sites where water level and snow height retrievals through SNR can be performed. In the sites, marked with *red dots* (dark gray in print version), no SNR GNSS-R retrievals of environmental parameters are possible. *(From Simeonov, T. (2021). Derivation and analysis of atmospheric water vapor and soil moisture from ground-based GNSS stations. In Technische Universität Berlin. Technische Universität Berlin. Global map of all IGS sites.)*

moisture, compared to the GNSS-R. This is due to the difference in the sampling area of the two methods (Simeonov, 2021).

In another study, the feasibility of using the International GNSS Service network for soil moisture retrievals is studied. The analysis shows that out of more than 500 stations, only around 30 stations can be used for reflectometry using the SNR observations methodology (see Fig. 8.20). The stations, where soil moisture observations have been successfully performed, cover all climate zones and continents. The soil moisture retrievals are highly dependent on local area conditions and the locations of the GNSS sites (Simeonov, 2021).

Spaceborne soil moisture monitoring with GNSS-R

The spaceborne soil moisture observations are carried on LEO satellites with GNSS-R receivers, described in the spaceborne wind measurement retrievals over oceans. Since the GNSS-R receivers can record reflections, coming not only from the sea surface but also from the ground, the reflections can be used to monitor properties of the land. Unlike the wind speed measurements, where DDMs are used, the spaceborne soil moisture monitoring uses the forward scattered signal power ($P_{r,\ eff}$). Studies have shown that the effective forward scattered signal power of wet surfaces produce stronger reflections than dry surfaces (Chew et al., 2014). Wetter surfaces have higher dielectric constants, which results in higher reflectivity than drier surfaces (Egido et al., 2012). Fig. 8.21 presents the $P_{r,\ eff}$ (left) and soil moisture VWC (right) for May (top), August (center), and the difference (bottom) (Chew & Small, 2018). The VWC measurements are from the Soil Moisture Active Passive (SMAP) satellite.

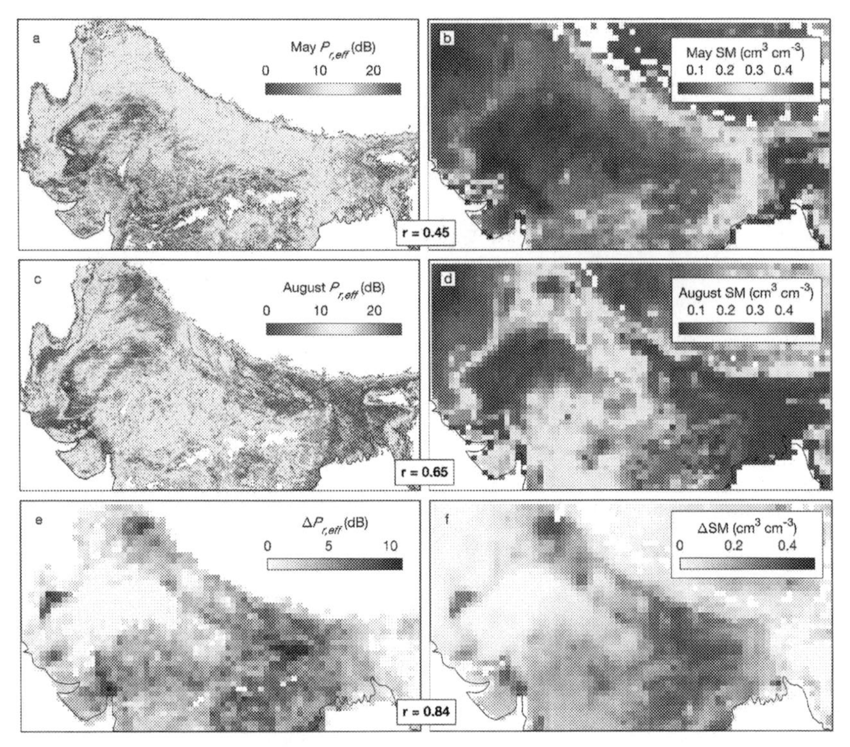

Fig. 8.21 India. *(Left)* Reflected power from the land, as recorded by the CYGNSS satellites constellation. *(Right)* The corresponding VWC measurements from the SMOS and SMAP satellites. *(Middle)* The spatial correlation between the two figures is recorded. *(From Chew, C., & Small, E. (2018). Soil moisture sensing using spaceborne GNSS reflections: Comparison of CYGNSS reflectivity to SMAP soil moisture. Geophysical Research Letters, 45(9), 4049–4057.)*

References

Asgarimehr, M., Wickert, J., & Reich, S. (2018). TDS-1 GNSS reflectometry: Development and validation of forward scattering winds. *IEEE Journal of Selected Topics in Applied Earth Observations and Remote Sensing*, *11*(11), 4534–4541.

Asgarimehr, M., Zavorotny, V., Wickert, J., & Reich, S. (2018). Can GNSS reflectometry detect precipitation over oceans? *Geophysical Research Letters*, *45*(22), 12–585.

Brady, N. C., & Weil, R. (2013). *Nature and properties of soils, the: Pearson new international edition*. Pearson Higher Ed.

Chew, C., & Small, E. (2018). Soil moisture sensing using spaceborne GNSS reflections: Comparison of CYGNSS reflectivity to SMAP soil moisture. *Geophysical Research Letters*, *45*(9), 4049–4057.

Chew, C. C., Small, E. E., Larson, K. M., & Zavorotny, V. U. (2014). Vegetation sensing using GPS-interferometric reflectometry: Theoretical effects of canopy parameters on signal-to-noise ratio data. *IEEE Transactions on Geoscience and Remote Sensing*, *53*(5), 2755–2764.

Egido, A., Caparrini, M., Ruffini, G., Paloscia, S., Santi, E., Guerriero, L., et al. (2012). Global navigation satellite systems reflectometry as a remote sensing tool for agriculture. *Remote Sensing*, *4*(8), 2356–2372.

Gleason, S., Hodgart, S., Sun, Y., Gommenginger, C., Mackin, S., Adjrad, M., et al. (2005). Detection and processing of bistatically reflected GPS signals from low earth orbit for the purpose of ocean remote sensing. *IEEE Transactions on Geoscience and Remote Sensing*, *43*(6), 1229–1241.

Jin, S., Cardellach, E., & Xie, F. (2014). *GNSS remote sensing*. Springer.

Larson, K. M., Small, E. E., Gutmann, E. D., Bilich, A. L., Braun, J. J., & Zavorotny, V. U. (2008). Use of GPS receivers as a soil moisture network for water cycle studies. *Geophysical Research Letters*, *35*(24).

Löfgren, J. S., & Haas, R. (2014). Sea level observations using multi-system GNSS reflectometry. In *NKG, 17th general assembly*.

Lowe, S. T., LaBrecque, J. L., Zuffada, C., Romans, L. J., Young, L. E., & Hajj, G. A. (2002). First spaceborne observation of an Earth-reflected GPS signal. *Radio Science*, *37*(1), 1–28.

Martin-Neira, M. (1993). A passive reflectometry and interferometry system (PARIS): Application to ocean altimetry. *ESA Journal*, *17*(4), 331–355.

Ruf, C. S., Chew, C., Lang, T., Morris, M. G., Nave, K., Ridley, A., et al. (2018). A new paradigm in earth environmental monitoring with the CYGNSS small satellite constellation. *Scientific Reports*, *8*(1), 1–13.

Semmling, A., Beckheinrich, J., Wickert, J., Beyerle, G., Schön, S., Fabra, F., et al. (2014). Sea surface topography retrieved from GNSS reflectometry phase data of the GEO-HALO flight mission. *Geophysical Research Letters*, *41*(3), 954–960.

Semmling, M., Schön, S., Beckheinrich, J., Beyerle, G., Ge, M., & Wickert, J. (2014). Carrier phase altimetry using Zeppelin based GNSS-R observations and water gauge reference data. In *EGUGA* (p. 11787).

Simeonov, T. (2021). Derivation and analysis of atmospheric water vapour and soil moisture from ground-based GNSS stations. In *Technische Universität Berlin* Technische Universität Berlin.

Zavorotny, V. U., & Voronovich, A. G. (2000). Scattering of GPS signals from the ocean with wind remote sensing application. *IEEE Transactions on Geoscience and Remote Sensing*, *38*(2), 951–964.

CHAPTER 9

Atmospheric data in GNSS processing

The main focus of this book is to show how atmospheric and environmental parameters can be measured, using the errors in the GNSS signal propagation. In this chapter, the roles of the observations are reversed. In certain applications, high accuracy of real-time GNSS positioning is required. Theoretically, the precise point positioning (PPP) and the double difference network (DGNSS) approaches of GNSS data processing are equivalent methods as it concerns the solution redundancy and estimated parameters. Practically, each strategy has advantages and disadvantages. While the DGNSS approach is a long-term traditional method, the PPP is considered as its modern alternative. Until now, the DGNSS approach has dominated the GNSS processing for tropospheric monitoring with latencies below 90 min after the observation (near real time). In 2013, the International GNSS Service (IGS) launched the real-time service (Caissy, Agrotis, Weber, Hernandez-Pajares, & Hugentobler, 2012), providing precise satellite orbit and clock corrections in support of ultrafast or real-time PPP processing. With the evolution of the IGS precise products and models the advantages of PPP became obvious: easy production in both real time and near real time, full flexibility for central or distributed processing, and the possibility of high spatiotemporal resolution estimates of advanced tropospheric parameters (zenith total delays—ZTD, horizontal gradients, and STDs). All these characteristics profit from a highly efficient and autonomous PPP processing strategy. However, to achieve the centimeter-level accuracy, PPP requires very accurate error estimation. Thus accurate error estimation for the tropospheric delays is required by the PPP processing strategy. Chapter 5 has demonstrated that NWP models can simulate with high degree of accuracy the amount of water vapor in the atmosphere. Thus instead of using GNSS delays to measure atmospheric phenomena, the NWP modeled data can be used to correct for signal delays in the atmosphere for higher accuracy of GNSS positioning.

Global Navigation Satellite System Monitoring of the Atmosphere
https://doi.org/10.1016/B978-0-12-819152-1.00003-9

GNSS mapping functions with atmospheric reanalysis/ NWP model

Historically, the use of atmospheric data as input in GNSS processing has been either for developing mapping functions that map the slant delays to zenith direction or for the separation of ZTD into a hydrostatic and wet delay. The first mapping functions using atmospheric reanalysis products are Vienna Mapping Function (VMF, Boehm, Kouba, & Schuh, 2009) and the Global Mapping Function (GMF, Böhm, Niell, Tregoning, & Schuh, 2006). Monthly profiles of atmospheric pressure, temperature, and water vapor pressure from the European Center for Medium-range Weather Forecast (ECMWF) atmospheric reanalysis (ERA-40) are used for the computation of the first VMF (VMF1). Third VMF (VMF3) release is available with both ERA-Interim reanalysis and ECMWF global operations NWP model. The temporal resolution of VMF3 is 6-hourly. The Technical University of Vienna operates the Vienna Mapping Functions Open Access Data service (VMF-service) providing a global next-day forecast of VMF3. A schematic representation of the VMF service with the available products is presented in Fig. 9.1. It is to be noted that atmospheric data input is used not only for GNSS but also for other space geodetic techniques (VLBI, DORIS, SLR).

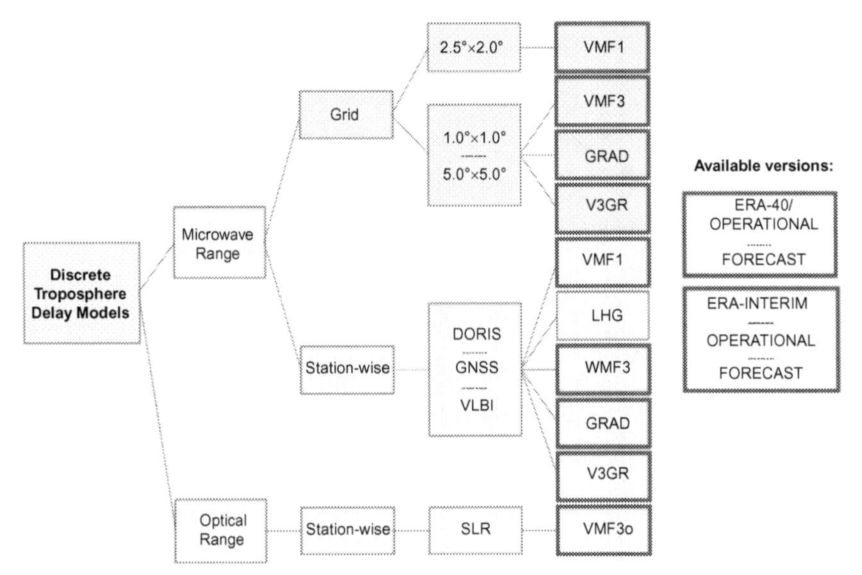

Fig. 9.1 VMF-service. Schematic representation of the VMF-service at Technical University of Vienna. The meteorological input is from ERA40/ERA-Interim reanalysis. *(From VMF Data Server: https://vmf.geo.tuwien.ac.at/products.html.)*

NWP data for GNSS Precise Point Positioning Processing (PPP)

Increased demand for accuracy for positioning and navigation and increasing resolution of NWP models resulted in the growing use of meteorological data in space-based geodetic techniques in the last decade. In particular, the PPP processing strategy is based on original observations or their linear combination without differencing between receivers or satellites. The pioneer in PPP processing strategy is the German Research Center for Geosciences (GFZ), which is providing PPP-based products since 2001 (Dick, Gendt, & Reigber, 2001; Gendt et al., 2004). However, this requires in-house processing of GNSS satellite clock and orbit products to enable the PPP processing for individual stations. A major step for the wide adoption of PPP strategy was the International GNSS Service (IGS) real-time service started in 2013 (Caissy et al., 2012). IGS real-time service provides precise satellite orbit and clock corrections in support of ultrafast or real-time PPP processing. The availability of global real-time data flow, software, and standards specified for precise product dissemination made PPP a viable approach for the subhourly products required by short-range weather forecasting of severe hail and thunderstorms, intense precipitation, etc. In Europe, several groups developed PPP processing using NWP data as an input. Here, three approaches developed in the Czech Republic, Poland, and UK are summarized.

PPP and NWP implementation at Geodetic Observatory Pecny

To benefit from the synergy between NWP and GNSS data the Geodetic Observatory Pecny, Czech Republic (GOP) developed a new concept based on a dual-layer tropospheric correction model for GNSS precise real-time positioning applications. The concept combines and predicts optimally GNSS hydrostatic and wet components and resembles the NWP assimilation but applied to GNSS tropospheric path delays of an electromagnetic signal with frequency up to 15 GHz. This new concept enhances the original GOP augmentation model introduced by Dousa, Vaclavovic, et al. (2015) and Dousa, Elias, et al. (2015) by combining NWP data with ZTDs estimated from GNSS permanent stations in regional networks. The first layer uses the NWP model forecast 01–48 h as GNSS background for ZHD and ZWD together with auxiliary parameters for the parameter vertical scaling. The second layer is an optimal combination of the background NWP model and GNSS near real-time tropospheric products. The optimum correction is achieved by using NWP for the hydrostatic delay modeling and for vertical scaling and GNSS NRT products for the nonhydrostatic delay (Douša, Eliaš, Václavovic, Eben, & Krč, 2018).

The concept is implemented in the "G-Nut/Shu" software. G-Nut/Shu software extracts tropospheric parameters from the NWP model field. The software is designed to support a flexible functionality for testing and assessing different variants of ZWD weighting method. In Fig. 9.2 the two-layer concept is presented. For the first layer two NWP models were used: the global reanalysis ERA-Interim (Dee et al., 2011) with 6-h temporal resolution and 1 degree horizontal resolution, as well as the mesoscale NWP Weather Research and Forecasting (WRF) with 1-h temporal resolution and 9-km horizontal resolution (Douša et al., 2020). The results demonstrate that the two-layer concept yields highly accurate and stable results. In particular, the second layer (bottom panel in Fig. 9.2) significantly improves the nonhydrostatic part of the NWP tropospheric background and also corrects possible errors in the hydrostatic NWP background. The most significant improvement (43%) of using the second layer is in terms of ZTD standard deviation. Douša et al. (2020) reported that the most important improvement of the combination of GNSS and NWP model is in terms of stability in the session-to-session performance of the dual-layer model ZTD statistics. The scatters in session-to-session mean standard deviations reduced from 8.7 to 4.4 mm and from 9.6 to 3.0 mm for the WRF model when using GNSS ZTD and ZTD together with horizontal gradients, respectively. G-Nut/Shu software is further developed to handle mesoscale models, such as WRF, Aladin, and Harmony, the global NWP model Global Forecasting System (GFS), and the global ERA-Interim reanalysis. In addition, it has other useful functionalities available at: https://www.pecny.cz/Joomla25/index.php/gnss/sw/shu.

PPP and NWP implementation at Wrocław University of Environmental and Life Sciences

Douša et al. (2020) presented the NWP integration in PPP processing implemented at Wrocław University of Environmental and Life Sciences, Republic of Poland. The PPP software Wroclaw Algorithms for real-time positioning (GNSS-WARP) uses NWP WRF model troposphere corrections over Poland. The model horizontal resolution is 4 km and has 47 vertical levels. To take advantage of the high spatiotemporal resolution the WRF model is also used to compute the mapping functions (WRFMF). To reconstruct the tropospheric delays the WRF model and

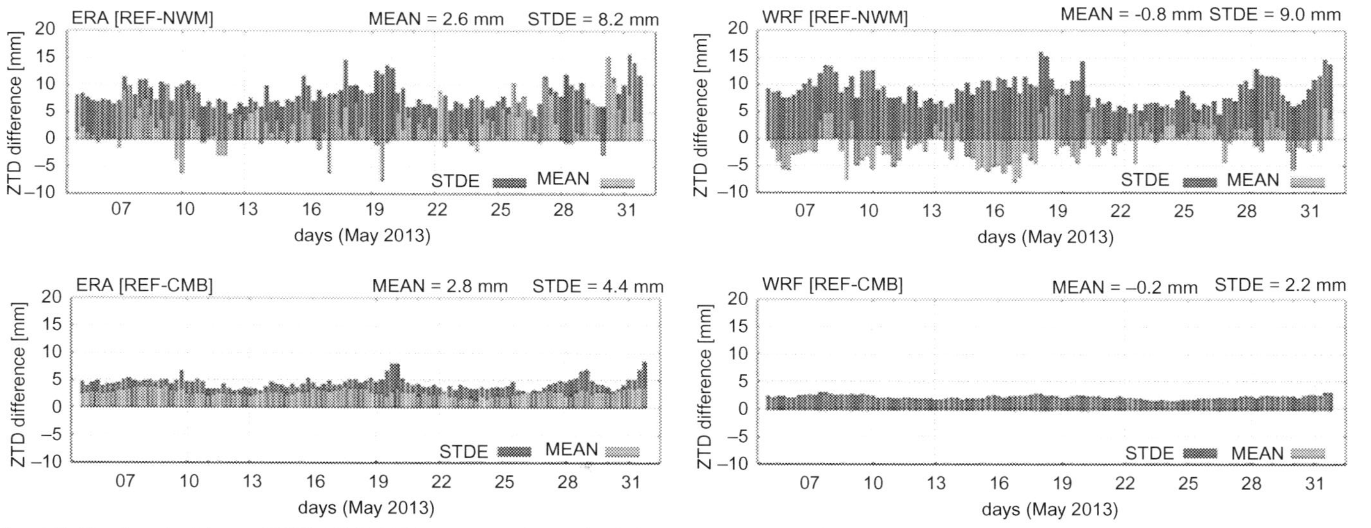

Fig. 9.2 Single-, dual-layer GNSS-NWP. ZTD mean and standard deviation statistics for (1) a single-layer NWP only *(top panel)* and (2) dual-layer model GNSS and NWP *(bottom panel)*. NWP input is from ERA-Interim *(left)* and WRF *(right)* for May 2013. *(From Jones, J., Guerova, G., Douša, J., Dick, G., de Haan, S., Pottiaux, E., & van Malderen, R. (2019). Advanced GNSS tropospheric products for monitoring severe weather events and climate. COST Action ES1206 Final Action Dissemination Report, 563.)*

near real-time GNSS data are combined using the least-squares collocation estimation. They calculated the North, East, and up coordinates of the 14 GNSS stations using the different tropospheric corrections. Static, continuous kinematic, and reinitialized kinematic coordinates are also estimated. The application of a high-resolution WRF-/GNSS-based ZTD model and mapping functions results in the best agreement with the official coordinates. In both static and kinematic mode, the application of high-resolution WRF and GNSS-based tropospheric corrections resulted in an average reduction of height bias by 20 and 10 mm, respectively. However, Douša et al. (2020) reported an increase of standard deviations by 1.5 and 4 mm for static and kinematic coordinates, respectively. Application of high-resolution WRF-/GNSS-based tropospheric corrections is found to shorten the convergence time for a 10-cm convergence level by 13% for the horizontal components and by 20% for the vertical component (Fig. 9.3). It is to be noted that only the tropospheric corrections based on NWP model are found to be biased in particular due to the model humidity field overestimation after rainfall (Wilgan, Hadas, Hordyniec, & Bosy, 2017).

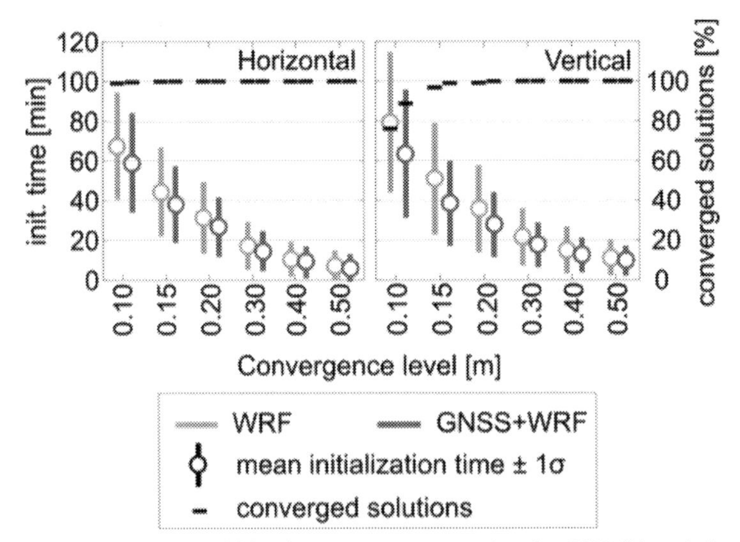

Fig. 9.3 PPP convergence. PPP solution convergence time for WRF *(blue; dark gray in print version, left bar from each pair)* and WRF with GNSS *(red; dark gray in print version, right bar from each pair)* for horizontal *(left subplot)* and vertical *(right subplot)* components. *(Reproduced from Wilgan, K., Hadas, T., Hordyniec, P., & Bosy, J. (2017). Real-time precise point positioning augmented with high-resolution numerical weather prediction model. GPS Solutions, 21(3), 1341–1353.)*

NWP augmented GNSS correction service for the UK

Douša et al. (2020) reported on PPP-based strategy for GNSS correction service. The GNSS correction service was developed by the University of Nottingham and one of the aims was to provide external corrections based on the Unified Model (UM) data produced by the UK Met Office (UKMO). The service is built as a five-tier system presented in Fig. 9.4. The tier 0 (ESA-B) is based on empirical ZTD estimations. The tier 1 (ESA-S) ZTD estimation uses surface meteorological parameters and improves the results by 27.3% in comparison to tier 0. The tier 2 ZTD estimation includes vertical lapse rates of meteorological parameters (ESA-A) and results in 45.2% improvement to tier 0. Tier 3 ZTD estimation is based on ZTD and mapping function correction and improves by 73.9% compared to tier 0. Douša et al. (2020) concluded that better ZTD estimation accuracy is obtained by more accurate atmospheric state information, i.e., the higher tier.

For tier 5 ray tracing the data processing is summarized in Fig. 9.5. From the NWP model (NWM) three-dimensional fields of pressure, temperature, and relative humidity (P, T, RH) the refractivity along the ray between satellite and ground-based station is calculated and then the tropospheric correction is obtained (D_{trop}^{NWM}).

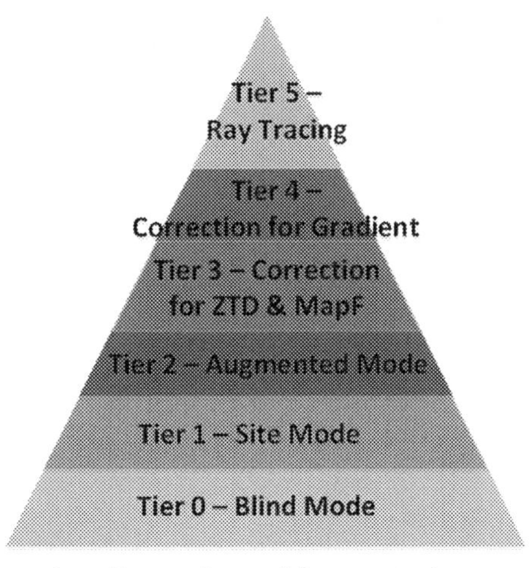

Fig. 9.4 UK GNSS service. Troposphere delay correction structure for UK GNSS correction service. *(From Jones, J., Guerova, G., Douša, J., Dick, G., de Haan, S., Pottiaux, E., & van Malderen, R. (2019). Advanced GNSS tropospheric products for monitoring severe weather events and climate. COST Action ES1206 Final Action Dissemination Report, 563.)*

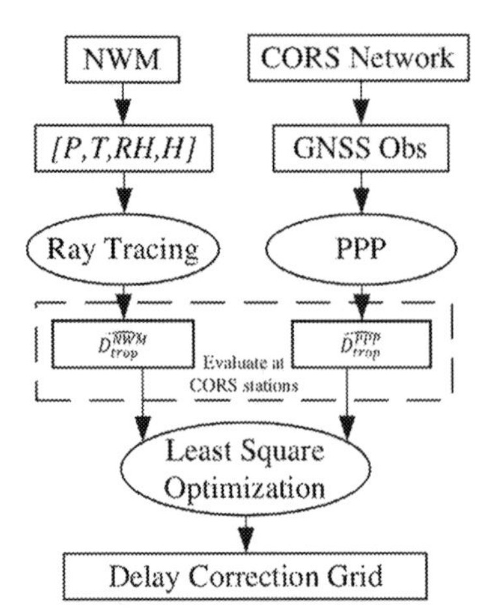

Fig. 9.5 NWP ray tracing. Data flow for GNSS STD and NWP ray tracing computation. *(From Yang, L., Hill, C., & Moore, T. (2013). Numerical weather modeling-based slant tropospheric delay estimation and its enhancement by GNSS data. Geo-spatial Information Science, 16 (3), 186–200.*

To test the contribution of each tier to the UK correction service a 12-month period is used. Fig. 9.6 presents the total occurrence probability of each tier for UK GNSS Continuously Operating Reference Stations (CORS) for elevation angles from 10 degree (top left) to 70 degree (bottom right). In each subfigure on the x-axis the accuracy threshold is in the range from 2 cm (left) to 10 cm (right). The y-axis gives the chance (in percentage) of each correction tier being required. At high elevations, a simple empirical estimation approach would be able to meet a lenient accuracy target. With an increasingly strict threshold, real information is required to replace empirical values to suppress the modeling error. Douša et al. (2020) concluded that for elevation angles below 25 degrees, the role of the tier 3 corrections (marked in black bars in Fig. 9.6) for low RMS residual thresholds decreases. Mapping functions (marked in green in Fig. 9.6, light gray above the black bars in print version) add value for elevation angles below 15 degrees, while gradients and ray-traced estimates (marked in red and yellow in Fig. 9.6, dark and gray bars in print version) are of importance for 10 degree elevation angle. As seen from the top panel of Fig. 9.6, the STD modeling assumptions based on the empirical mapping function and balanced local atmosphere profile gradually become less valid, and the error they bring cannot be ignored. This study demonstrates the required complexity of scouting the atmospheric error for a commercial GNSS correction service and can be used in setting up the next-generation service.

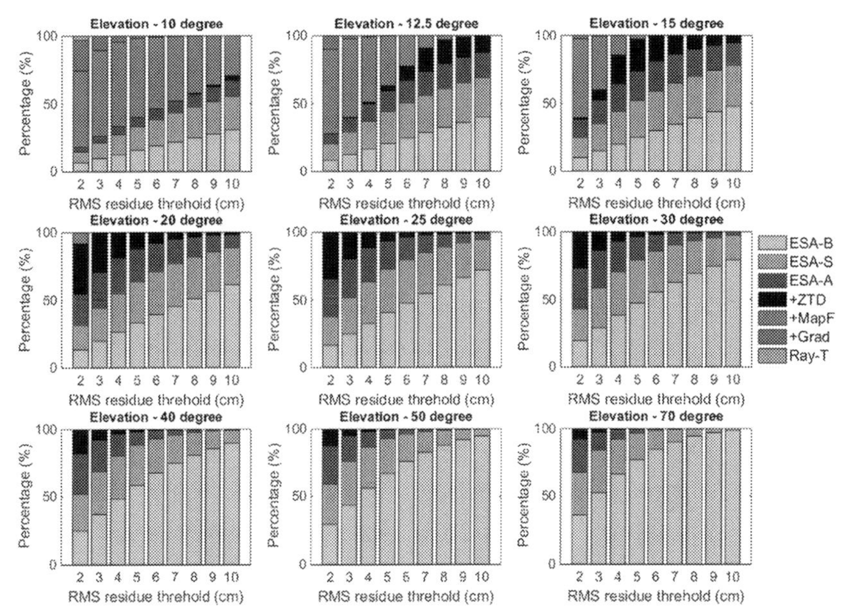

Fig. 9.6 Correction model contribution. Contribution in percentage for each correction model as a function of elevation angle. The higher the elevation angle, the more accurate the mapping function is required. The color scheme matches the one in Fig. 9.5 with the exception of tier 3 where the ZTD is given with black color. *(From Jones, J., Guerova, G., Douša, J., Dick, G., de Haan, S., Pottiaux, E., & van Malderen, R. (2019). Advanced GNSS tropospheric products for monitoring severe weather events and climate. COST Action ES1206 Final Action Dissemination Report, 563.)*

References

Boehm, J., Kouba, J., & Schuh, H. (2009). Forecast Vienna Mapping Functions 1 for real-time analysis of space geodetic observations. *Journal of Geodesy*, *83*(5), 397–401.

Böhm, J., Niell, A., Tregoning, P., & Schuh, H. (2006). Global Mapping Function (GMF): A new empirical mapping function based on numerical weather model data. *Geophysical Research Letters*, *33*(7).

Caissy, M., Agrotis, L., Weber, G., Hernandez-Pajares, M., & Hugentobler, U. (2012). The international GNSS real-time service. *GPS World*, *6*(23), 52–58.

Dee, D. P., Uppala, S. M., Simmons, A., Berrisford, P., Poli, P., Kobayashi, S., et al. (2011). The ERA-interim reanalysis: Configuration and performance of the data assimilation system. *Quarterly Journal of the Royal Meteorological Society*, *137*(656), 553–597.

Dick, G., Gendt, G., & Reigber, C. (2001). First experience with near real-time water vapor estimation in a German GPS network. *Journal of Atmospheric and Solar-Terrestrial Physics*, *63*(12), 1295–1304.

Douša, J., Dick, G., Altiner, Y., Alshawaf, F., Bosy, J., Brenot, H., et al. (2020). Advanced GNSS processing techniques (working group 1). In *Advanced GNSS tropospheric products for monitoring severe weather events and climate* (pp. 33–210). Springer.

Dousa, J., Vaclavovic, P., Krc, P., Elias, M., Eben, E., & Resler, J. (2015). NWM forecast monitoring with near real-time GNSS products. In *Proceedings of the 5th scientific Galileo colloquium*. Germany: Braunschweig.

Douša, J., Eliaš, M., Václavovic, P., Eben, K., & Krč, P. (2018). A two-stage tropospheric correction model combining data from GNSS and numerical weather model. *GPS Solutions*, *22*(3), 77. https://doi.org/10.1007/s10291-018-0742-x.

Dousa, J., Elias, M., Veerman, H., van Leeuwen, S. S., Zelle, H., de Haan, S., et al. (2015). High accuracy tropospheric delay determination based on improved modelling and high resolution Numerical Weather Model. In *Proceedings of the 28th international technical meeting of the satellite division of the institute of navigation (ION GNSS + 2015)* (pp. 3734–3744).

Gendt, G., Dick, G., Reigber, C., TOMASSINI, M., LIU, Y., & Ramatschi, M. (2004). Near real time GPS water vapor monitoring for numerical weather prediction in Germany. *Journal of the Meteorological Society of Japan Ser II*, *82*(1B), 361–370.

Wilgan, K., Hadas, T., Hordyniec, P., & Bosy, J. (2017). Real-time precise point positioning augmented with high-resolution numerical weather prediction model. *GPS Solutions*, *21*(3), 1341–1353.

Index

Note: Page numbers followed by *f* indicate figures and *t* indicate tables.

R

S

T

Printed in the United States
by Baker & Taylor Publisher Services